Springer Finance

Springer

London
Berlin
Heidelberg
New York
Barcelona
Hong Kong
Milan
Paris
Singapore
Tokyo

Springer Finance

Risk-Neutral Valuation: Pricing and Hedging of Financial Derivatives
N.H Bingham and Rüdiger Kiesel
ISBN 1-85233-001-5 (1998)

Visual Explorations in Finance with Self-Organizing Maps
Guido Deboeck and Teuvo Kohonen (Eds)
ISBN 3-540-76266-3 (1998)

Mathematics of Financial Markets
Robert J. Elliott and P. Ekkehard Kopp
ISBN 0-387-98553-0 (1998)

Mathematical Models of Financial Derivatives
Y.-K. Kwok
ISBN 981-3083-25-5 (1998)

Antoon Pelsser

Efficient Methods for Valuing Interest Rate Derivatives

 Springer

Antoon Pelsser, PhD
Erasmus University Rotterdam, Department of Finance, PO Box 1738,
3000 DR Rotterdam, The Netherlands • *email* pelsser@few.eur.nl

Springer London Series Advisors

Professor Giovanni Barone-Adesi, University of Alberta, Canada
Dr Ekkehard Kopp, University of Hull, UK

The models discussed in this book present an overview of the academic literature on interest rate derivative modelling. The models discussed here are not a reflection of the models in use by ABN-Amro Bank N.V. at the time of writing this manuscript.

ISBN 1-85233-304-9 Springer-Verlag London Berlin Heidelberg

British Library Cataloguing in Publication Data
Pelsser, Antoon
  Efficient methods for valuing interest rate derivatives. -
  (Springer finance)
  1. Interest rate futures – Mathematical models 2. Derivative
  securities – Mathematical models
  I. Title
  332.6'323'015118
ISBN 1852333049

Library of Congress Cataloging-in-Publication Data
Pelsser, Antoon, 1968-
  Efficient methods for valuing interest rate derivatives / Antoon Pelsser.
    p.   cm. -- (Springer finance)
  Includes bibliographical references and index.
  ISBN 1-85233-304-9 (alk. paper)
  1. Derivative securities. I. Title. II. Series.
HG6024.A3 P45   2000
332.63'23--dc21   00-033821

Typesetting: Camera-ready by author
Printed and bound at the Athenæum Press Ltd., Gateshead, Tyne & Wear
12/3830-543210 Printed on acid-free paper SPIN 10763773

This book is dedicated to the memory of my father André Pelsser.

'Ich han dich noe wal laank,
hei ich dich noe nog mèr groët.'

Preface

This book aims to give an overview of models that can be used for efficient valuation of (exotic) interest rate derivatives. The first part of this book discusses and compares traditional models such as spot and forward rate models which are widely used both by academics and practitioners. The second part of this book focuses on models that have been developed recently: the market models. These models have already sparked a lot of interest with banks and institutions. However, since the underlying mathematics is more complicated, it can be difficult to understand and implement these models successfully. This book seeks to de-mystify the market models and aims to show how these models can be implemented by using examples of products that are actually traded in the market. Also we discuss how to choose the model most suited to different products and we show that for many popular products a simple modelling approach based on convexity correction is very successful.

The book is aimed at people with a solid quantitative background looking for a good guidebook to interest rate derivative modelling, such as quantitative researchers, risk managers, risk controllers or (exotic) interest rate derivative traders. I use the term "guidebook" deliberately as this book reflects very much my own experience and reflects my personal views on how to value interest rate derivatives adequately and how to avoid common pitfalls.

The first part of this book has grown out of my PhD thesis at the Erasmus University in Rotterdam. Winning the Christiaan Huygens price in October 1999 (which is awarded by the Royal Dutch Academy of Sciences for the best thesis written in the area of Econometrics and Actuarial Science of the past 4 years) inspired me to rewrite my thesis into a book and to add several new chapters discussing the most recent developments in the area.

During the period I was working on my thesis and working as a "quant" I received help and guidance from a lot of different people. I would especially like to thank Ton Vorst, Douglas Bongartz-Renaud and Phil Hunt. Ton lured me, by asking me to be his research assistant, into the world of option pricing and has taught me how to be a scientist. Douglas lured me, by hiring me for ABN-Amro, into the hectic environment of working in a dealing room and has taught me how to survive there. Phil and I have worked several years closely together. Phil lured me into the world of martingales and probability measures and has made me appreciate the power of these tools in the context of derivatives modelling. We have spent many a joyful hour together solving modelling problems.

Furthermore, I would like to thank Frank de Jong, Joost Driessen, Joanne Kennedy and Juan Moraleda. With each of them, I have in recent years co-authored scientific papers. I would also like to thank my colleagues at ABN-Amro Bank: Richard Averill, Jelle Beenen, Pauline Bod and Mark de Vries. They read various draft versions and gave valuable comments and suggestions. And I would like to thank Karen Barker, the Editorial Assistant at Springer Verlag, for guiding me through the book preparation process.

Last, but not least, I would like to thank my wife Chantal for loving and supporting me; and for putting me gently back on track when I spend too much time on mathematics. . .

Amsterdam, March 2000   Antoon Pelsser

The author busy explaining a model.

Contents

1. **Introduction** 1

2. **Arbitrage, Martingales and Numerical Methods** 5

    2.1   Arbitrage and Martingales 6
         2.1.1   Basic Setup 6
         2.1.2   Equivalent Martingale Measure 8
         2.1.3   Change of Numeraire Theorem 10
         2.1.4   Girsanov's Theorem and Itô's Lemma 11
         2.1.5   Application: Black-Scholes Model 12
         2.1.6   Application: Foreign-Exchange Options 14
    2.2   Numerical Methods 16
         2.2.1   Derivation of Black-Scholes Partial Differential
                Equation 16
         2.2.2   Feynman-Kac Formula 17
         2.2.3   Numerical Solution of PDE's 18
         2.2.4   Monte Carlo Simulation 18
         2.2.5   Numerical Integration 20

Part I. Spot and Forward Rate Models

3. **Spot and Forward Rate Models** 23

    3.1   Vasicek Methodology 23
         3.1.1   Spot Interest Rate 23
         3.1.2   Partial Differential Equation 24
         3.1.3   Calculating Prices 25
         3.1.4   Example: Ho-Lee Model 26
    3.2   Heath-Jarrow-Morton Methodology 27
         3.2.1   Forward Rates 27
         3.2.2   Equivalent Martingale Measure 28
         3.2.3   Calculating Prices 29
         3.2.4   Example: Ho-Lee Model 30
    3.3   Equivalence of the Methodologies 30

4. Fundamental Solutions and the Forward-Risk-Adjusted Measure 31

    4.1   Forward-Risk-Adjusted Measure 32
    4.2   Fundamental Solutions 34
    4.3   Obtaining Fundamental Solutions 36
    4.4   Example: Ho-Lee Model 37
          4.4.1   Radon-Nikodym Derivative 37
          4.4.2   Fundamental Solutions 38
    4.5   Fundamental Solutions for Normal Models 40

5. The Hull-White Model 45

    5.1   Spot Rate Process 46
          5.1.1   Partial Differential Equation 47
          5.1.2   Transformation of Variables 47
    5.2   Analytical Formulæ 48
          5.2.1   Fundamental Solutions 49
          5.2.2   Option Prices 50
          5.2.3   Prices for Other Instruments 51
    5.3   Implementation of the Model 52
          5.3.1   Fitting the Model to the Initial Term-Structure 52
          5.3.2   Transformation of Variables 53
          5.3.3   Trinomial Tree 53
    5.4   Performance of the Algorithm 55
    5.5   Appendix 57

6. The Squared Gaussian Model 59

    6.1   Spot Rate Process 60
          6.1.1   Partial Differential Equation 60
    6.2   Analytical Formulæ 61
          6.2.1   Fundamental Solutions 62
          6.2.2   Option Prices 63
    6.3   Implementation of the Model 64
          6.3.1   Fitting the Model to the Initial Term-Structure 64
          6.3.2   Trinomial Tree 66
    6.4   Appendix A 66
    6.5   Appendix B 69

7. An Empirical Comparison of One-Factor Models 71

    7.1   Yield-Curve Models 72
    7.2   Econometric Approach 74
    7.3   Data 77
    7.4   Empirical Results 77
    7.5   Conclusions 84

Part II. Market Rate Models

8. LIBOR and Swap Market Models 87
   8.1 LIBOR Market Models 88
      8.1.1 LIBOR Process 88
      8.1.2 Caplet Price 89
      8.1.3 Terminal Measure 90
   8.2 Swap Market Models 91
      8.2.1 Interest Rate Swaps 92
      8.2.2 Swaption Price 93
      8.2.3 Terminal Measure 95
      8.2.4 T_1-Forward Measure 96
   8.3 Monte Carlo Simulation for LIBOR Market Models 97
      8.3.1 Calculating the Numeraire Rebased Payoff 98
      8.3.2 Example: Vanilla Cap 99
      8.3.3 Discrete Barrier Caps/Floors 100
      8.3.4 Discrete Barrier Digital Caps/Floors 102
      8.3.5 Payment Stream 103
      8.3.6 Ratchets 103
   8.4 Monte Carlo Simulation for Swap Market Models 104
      8.4.1 Terminal Measure 104
      8.4.2 T_1-Forward Measure 105
      8.4.3 Example: Spread Option 106

9. Markov-Functional Models 109
   9.1 Basic Assumptions 110
   9.2 LIBOR Markov-Functional Model 111
   9.3 Swap Markov-Functional Model 114
   9.4 Numerical Implementation 115
      9.4.1 Numerical Integration 115
      9.4.2 Non-Parametric Implementation 117
      9.4.3 Semi-Parametric Implementation 118
   9.5 Forward Volatilities and Auto-Correlation 120
      9.5.1 Mean-Reversion and Auto-Correlation 120
      9.5.2 Auto-Correlation and the Volatility Function 121
   9.6 LIBOR Example: Barrier Caps 121
      9.6.1 Numerical Calculation 121
      9.6.2 Comparison with LIBOR Market Model 123
      9.6.3 Impact of Mean-Reversion 124
   9.7 LIBOR Example: Chooser- and Auto-Caps 125
      9.7.1 Auto-Caps/Floors 125
      9.7.2 Chooser-Caps/Floors 125
      9.7.3 Auto- and Chooser-Digitals 125
      9.7.4 Numerical Implementation 125

9.8   Swap Example: Bermudan Swaptions 127
      9.8.1   Early Notification 127
      9.8.2   Comparison Between Models 128

10. An Empirical Comparison of Market Models 131
      10.1 Data Description 132
      10.2 LIBOR Market Model 132
           10.2.1 Calibration Methodology 132
           10.2.2 Estimation and Pricing Results 134
      10.3 Swap Market Model 135
           10.3.1 Calibration Methodology 135
           10.3.2 Estimation and Pricing Results 135
      10.4 Conclusion 136

11. Convexity Correction 139
      11.1 Convexity Correction and Change of Numeraire 140
           11.1.1 Multi-Currency Change of Numeraire Theorem 140
           11.1.2 Convexity Correction 142
      11.2 Options on Convexity Corrected Rates 145
           11.2.1 Option Price Formula 146
           11.2.2 Digital Price Formula 147
      11.3 Single Index Products 147
           11.3.1 LIBOR in Arrears 147
           11.3.2 Constant Maturity Swap 149
           11.3.3 Diffed LIBOR 150
           11.3.4 Diffed CMS 150
      11.4 Multi-Index Products 151
           11.4.1 Rate Based Spread Options 151
           11.4.2 Spread Digital 153
           11.4.3 Other Multi-Index Products 153
           11.4.4 Comparison with Market Models 154
      11.5 A Warning on Convexity Correction 155
      11.6 Appendix: Linear Swap Rate Model 156

12. Extensions and Further Developments 159
      12.1 General Philosophy 159
      12.2 Multi-Factor Models 160
      12.3 Volatility Skews 161

References 163

Index 167

1. Introduction

Since the opening of the first options exchange in Chicago in 1973, the financial world has witnessed an explosive growth in the trading of derivative securities. Since that time, exchanges where futures and options can be traded have been opened all over the world and the volume of contracts traded worldwide has grown enormously.

The growth in derivatives markets has not only been a growth in volume, but also a growth in complexity. Most of these more complex derivative contracts are not exchange traded, but are traded "over-the-counter". Often, the over-the-counter contracts are created by banks to provide tailor made products to reduce financial risks for clients. In this respect, banks are playing an innovative and important role in providing a market for the exchange of financial risks.

However, derivatives can also be used to create highly leveraged speculative positions, which can lead to large profits, or large losses. In recent years, several companies and hedge funds have suffered large losses due to speculative trading in derivatives. In some cases these losses made the headlines of the financial press. This has created a general feeling that "derivatives are dangerous". Some people have called for a strict regulation of derivatives markets, or even for a complete ban of over-the-counter trading.

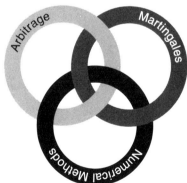

Although these claims are an overreaction to what has happened recently, it has become apparent that it is important, both for market participants and regulators, to have a good insight into the pricing and risk characteristics of derivatives. Considerable academic research has been devoted to the valuation of derivative securities. Since the seminal research of Fisher Black, Myron Scholes and Robert Merton in the early seventies, an elegant theory has been developed.

The figure above (and also the cover of this book) shows three interconnected rings which are labelled arbitrage, martingales and numerical methods. An arbitrage opportunity offers the opportunity of generating wealth at some future time without an initial investment. Due to competitive forces, it seems reasonable that arbitrage opportunities cannot exist in an economy that is in equilibrium. This is often summarised by the phrase "there is no such thing as a free lunch." A martingale is the mathematical formalisation of the concept of a fair game. If prices of derivative securities can be modelled as martingales this implies that no market participant can consistently make (or lose) money by trading in derivative securities. Numerical methods will provide us with the tools to explicitly calculate prices of many derivatives for which no analytical expressions can be found. Since we would like to consider models that are useful in a trading context, we seek models which permit efficient numerical valuation methods for exotic interest rate derivatives.

The concepts of arbitrage, martingales and numerical methods are the three pillars on which the theory and implementation of the valuation of derivative securities rests. A closer inspection of the rings will reveal that every pair of rings is not connected, only the three rings together are connected. The same is true for the theory of valuing derivative securities. A good command of the concepts of arbitrage, martingales and numerical methods is needed to obtain a coherent understanding of the valuation of derivative securities. Chapter 2, which explains the basic theory of the valuation of derivatives is therefore called 'Arbitrage, Martingales and Numerical Methods'.

As was argued before, the markets in derivative securities can be viewed as insurance markets for financial risks. Since the Fed decided in 1979 to change its monetary policy, interest rate volatilities in the US have risen considerably. Due to the increasing globalisation of capital markets, this has led to an increase in interest rate volatilities world-wide. Many companies have sought to buy insurance against the increased uncertainty in interest rate markets. For this reason, the market for interest rate derivatives has been one of the fastest growing markets in the last two decades. Strong interest in this area has inspired a lot of research into modelling the behaviour of interest rates and the pricing and risk characteristics of interest rate derivatives.

In this book we will focus on models that can be used to value interest rate derivatives. To determine the prices of exotic interest rate derivatives, pricing models are used as an "extrapolation tool". Given the prices of liquid instruments available in the market, pricing models for exotic derivatives try to extract from these prices information about the probability distribution of the future values of the underlying assets and use this information to price exotic derivatives. In the case of interest rate derivatives, the liquid instruments are caps/floors or swaptions. In the markets, prices of these liquid instruments are readily available. To successfully extract information from the prices (or the implied volatilities) quoted in the markets we seek internally consistent models that can explain observed prices for the underlying instruments.

Two major types of modelling approaches can be distinguished. The first approach is to take instantaneous interest rates as the basis for modelling the term-structure of interest rates. An instantaneous spot interest rate is the interest one earns on a riskless investment over an infinitesimal time-period dt. Once we have a model for the evolution of the instantaneous spot interest rates, all other interest rates can be derived by integrating over the spot interest rates. Hence, these models are called spot rate models. This mathematically convenient choice leads to models which are particularly tractable. However, since these models are set up in terms of a mathematically convenient rate that does not exist in practice, valuation formulæ for real-world instruments like caps, floors and swaptions tend to be fairly complicated. To fit the spot rate models to the prices of these instruments we need complicated numerical procedures and the empirical results are not always satisfactory. We will focus our attention to models of this type in the first part of this book.

This shortcoming has inspired the second type of approach, which takes real market interest rates, like LIBOR or swap rates, as a basis for modelling. These models, which have emerged recently, are called market rate models. They tend to be more complicated in their setup, but the big advantage is that market standard pricing formulæ for the standard instruments can be reproduced with these models. Hence, by construction, these models can be made to fit the market prices perfectly. We will focus our attention to models of this type in the second part of this book. Also we will show how many exotic interest rate derivatives can be valued efficiently using market models.

The first part of this book is organised as follows. In Chapter 3 we show how spot interest rate derivatives can be valued theoretically. Interest rates play a double role when valuing interest rate derivatives, as they determine both the discounting and the payoff of the derivative. This makes interest rate derivatives more difficult to value. Hence, we explore in Chapter 4 some analytical methods that can be used to simplify the pricing of interest rate derivatives. In Chapters 5 and 6 we analyse two interest rate models. Due to the assumption that there is only one underlying source of uncertainty that drives the evolution of the interest rates, these models are relatively simple to understand and to analyse. We derive analytical formulæ for valuing interest rate derivatives, and we derive numerical methods to approximate the prices of derivatives for which no analytical pricing formulæ can be found. Finally, we make an empirical comparison of several one-factor spot rate models in Chapter 7. Unfortunately, it turns out that the models with a rich analytical structure do not describe the prices of interest rate derivatives very well.

The second part of this book focuses therefore on the market models, which have the advantage that they can be made to fit market prices perfectly. This second part is organised as follows. In Chapter 8 we describe LIBOR and swap market models, which was the first class of market models described in the literature. When these models were introduced it was shown for the first time that the use of Black's formula for the pricing of interest rate derivatives like caps/floors and swaptions is consistent with an arbitrage-free interest rate model. Because these models are consistent with the market standard Black formula, the implied volatilities quoted in the market to price standard interest rate derivatives can be used directly as an input for the market models, which greatly simplifies their calibration. To calculate prices for exotic interest rate derivatives, one has to use Monte Carlo simulation. We illustrate how Monte Carlo simulation procedures can be implemented to price path-dependent interest rate derivatives. For other types of exotic interest rate derivatives having American-style features, the use of Monte Carlo simulation is very cumbersome. Hence American-style interest rate derivatives cannot be valued efficiently with the market models of Chapter 8. To this end, we introduce in Chapter 9 the class of Markov-Functional models. This class of models combines the attractive feature of the market models (perfect fit to the market) with the attractive feature of the spot rate models (easy numerical implementation). We illustrate the use of Markov-Functional for various interest rate derivatives which possess American-style features. A problem with the use of (one-factor) market models is that one can either work with a LIBOR based model or work with a swap based model, but the two modelling approaches are mutually inconsistent. In Chapter 10, we therefore investigate empirically which of the two approaches seems to fit the behaviour of market prices best. From the preliminary investigations presented, it seems that using LIBOR rates as a basis for modelling provides the better description.

In Chapter 11 we digress a little from using term-structure models. Many exotic interest rate derivatives depend only on one or two interest rates. We show that we can successfully derive efficient valuation models by focussing our modelling efforts only on the rates under consideration. We show that for many different products, the exotic nature of the product can be captured by pricing the product as a vanilla interest rate product on an adjusted forward interest rate. This adjustment to the forward rate is known in the market as "convexity correction".

The final chapter is devoted to extensions and further developments of the models described in this book.

2. Arbitrage, Martingales and Numerical Methods

The cornerstone of option pricing theory is the assumption that any financial instrument which has a guaranteed non-negative payout must have a non-negative price. The existence of an instrument which would have non-negative payoffs and a negative price is called an *arbitrage opportunity*. If arbitrage opportunities would exist, it would be a means for investors to generate money without any initial investment. Of course, many investors would try to exploit the arbitrage opportunity, and due to the increased demand, the price would rise and the arbitrage opportunity would disappear. Hence, in an economy that is in equilibrium it seems reasonable to rule out the existence of arbitrage opportunities. Although the assumption that arbitrage opportunities do not exist seems a rather plausible and trivial assumption, we shall see it is indeed the foundation for all of the option pricing theory.

Another important assumption needed to get the edifice of option pricing off the ground is the *absence of transaction costs*. This means that assets can be bought and sold in the market for the same price. This assumption is clearly violated in real markets. In the presence of transaction costs, not all arbitrage opportunities which would theoretically be profitable can be exploited. However, large market participants (like banks and institutions) face very little transaction costs. These large players have the opportunity to exploit almost all arbitrage opportunities with large amounts of money and markets will be driven to an equilibrium close to the equilibrium that would prevail if transaction costs were absent. Hence, if we consider markets as a whole, the assumption that transaction costs are absent is a good approximation of the real world situation.

The rest of this chapter is divided in two sections. In the first section we provide the basic mathematical setup that will be used throughout this book. We furthermore show under which conditions an economy is free of arbitrage opportunities and how prices of derivative securities can be calculated. As an example we analyse the Black-Scholes model. In the second section we analyse analytical and numerical methods for solving option price models.

2.1 Arbitrage and Martingales

In this section we provide the basic mathematical setting in which the theory of option pricing can be cast. Without any proofs we summarise the key results that an economy is free of arbitrage opportunities if a probability measure can be found such that the prices of marketed assets become martingales. By setting up trading strategies which replicate the payoff of derivative securities, the martingale property can then be shown to carry over from the marketed assets to the prices of all derivative securities. Hence, the prices of all derivatives become martingales and this property can then be used to calculate prices for derivative securities. For readers interested in a more formal and rigorous treatment of these results, we refer to the books of Musiela and Rutkowski (1997), Karatzas and Shreve (1998) or Hunt and Kennedy (2000). For an excellent intuitive introduction we refer to Baxter and Rennie (1996).

2.1.1 Basic Setup

Throughout this book we consider a continuous trading economy, with a finite trading interval given by $[0, T]$. The uncertainty is modelled by the probability space $(\Omega, \mathcal{F}, \mathbb{P})$. In this notation, Ω denotes a sample space, with elements $\omega \in \Omega$; $\mathcal{F}$ denotes a σ-algebra on Ω; and $\mathbb{P}$ denotes a probability measure on $(\Omega, \mathcal{F})$ The uncertainty is resolved over $[0, T]$ according to a filtration $\{\mathcal{F}_t\}$ satisfying "the usual conditions".

Throughout this book we assume that there exist assets which are traded in a market. The assets are called *marketed assets*. We also assume that the prices $Z(t)$ of these marketed assets can be modelled via Itô processes which are described by stochastic differential equations

$$dZ(t) = \mu(t, \omega)\, dt + \sigma(t, \omega)\, dW, \tag{2.1}$$

where the functions $\mu(t, \omega)$ and $\sigma(t, \omega)$ are assumed to be $\mathcal{F}_t$-adapted and also satisfy

$$\int_0^T |\mu(t, \omega)|\, dt < \infty$$
$$\int_0^T \sigma(t, \omega)^2\, dt < \infty, \tag{2.2}$$

with probability one.

The observant reader may note that there is only one source of uncertainty (the Brownian motion W) that drives the prices of the marketed assets. It is relatively straightforward to set up the economy such that more sources of uncertainty drive the prices of the marketed assets, but the notation gets more complicated and would distract from the basic ideas we try to convey. Furthermore, for the largest part of this book we will be concerned with economies which are assumed to have only one source of uncertainty. However, in Chapters 8 and 11 we give examples of multi-factor models.

It is also true that the prices of marketed assets defined in (2.1) are less general than usual in the literature. The sample paths of Itô processes are continuous, which excludes, for example, discrete dividend payments. For generalisations along this direction, we refer to the literature overview in Karatzas and Shreve (1998) Sect. 1.8. However, in this book we will nowhere encounter marketed assets with discontinuous sample paths.

Suppose there are N marketed assets with prices $Z_1(t), \ldots, Z_N(t)$, which all follow Itô processes. A *trading strategy* is a predictable N-dimensional stochastic process $\delta(t, \omega) = (\delta_1(t, \omega), \ldots, \delta_N(t, \omega))$, where $\delta_n(t, \omega)$ denotes the holdings in asset n at time t. The asset holdings $\delta_n(t, \omega)$ are furthermore assumed to satisfy additional regularity conditions to which we will return later.

The value $V(\delta, t)$ at time t of a trading strategy δ is given by

$$V(\delta, t) = \sum_{n=1}^{N} \delta_n(t) Z_n(t). \tag{2.3}$$

A *self-financing trading strategy* is a strategy δ with the property

$$V(\delta, t) = V(\delta, 0) + \sum_{n=1}^{N} \int_0^t \delta_n(s) \, dZ_n(s), \quad \forall t \in [0, T], \tag{2.4}$$

where the integrals $\int \delta_n(s) \, dZ_n(s)$ denote Itô integrals. Hence, a self-financing trading strategy is a trading strategy that requires nor generates funds between time 0 and time T. Note, that in the definition of a self-financing trading strategy we have also included the modelling assumption that the gains from trading can be modelled as Itô integrals.

An *arbitrage opportunity* is a self-financing trading strategy δ, with $\Pr[V(\delta, T) \geq 0] = 1$ and $V(\delta, 0) < 0$. Hence, an arbitrage opportunity is a self-financing trading strategy which has strictly negative initial costs, and with probability one has a non-negative value at time T.

A *derivative security* is defined as a $\mathcal{F}_T$-measurable random variable $H(T)$. The random variable has to satisfy an additional regularity constraint to which we will return later. The random variable $H(T)$ can be interpreted as the (uncertain) payoff of the derivative security at time T. If we can find a self-financing trading strategy δ such that $V(\delta, T) = H(T)$ with probability one, the derivative is said to be *attainable*. The self-financing trading strategy is then called a *replicating strategy*. If in an economy all derivative securities are attainable, the economy is called *complete*.

If no arbitrage opportunities and no transaction costs exist in an economy, the value of a replicating strategy at time t gives a unique value for the attainable derivative $H(T)$. This is true, since (in the absence of transaction costs) the existence of two replicating strategies of the same derivative with different values would immediately create an arbitrage opportunity. Hence, we can determine the value of derivative securities by the value of the replicating portfolios. This is called *pricing by arbitrage*.

However, this raises two questions. First, under which conditions is a continuous trading economy free of arbitrage opportunities? Second, under which conditions is the economy complete? If these two conditions are satisfied, all derivative securities can be priced by arbitrage.

2.1.2 Equivalent Martingale Measure

The questions of no-arbitrage and completeness were first addressed mathematically in the seminal papers of Harrison and Kreps (1979) and Harrison and Pliska (1981). They showed that both questions can be solved at once using the notion of a martingale measure.

Any asset which has strictly positive prices for all $t \in [0,T]$ is called a *numeraire*. We can use numeraires to denominate all prices in an economy. Suppose that the marketed asset Z_1 is a numeraire. The prices of other marketed assets denominated in Z_1 are called the *relative prices* denoted by $Z_n' = Z_n/Z_1$.

Let $(\Omega, \mathcal{F}, \mathbb{P})$ denote the probability space from the previous subsection. Consider now the set that contains all probability measures $\mathbb{Q}^*$ such that:

i $\mathbb{Q}^*$ is equivalent to $\mathbb{P}$, i.e. both measures have the same null-sets;
ii the relative price processes Z_n' are martingales under $\mathbb{Q}^*$ for all n, i.e. for $t \leq s$ we have $\mathbb{E}^* \left(Z_n'(s) \mid \mathcal{F}_t \right) = Z_n'(t)$.

The measures $\mathbb{Q}^*$ are called *equivalent martingale measures*. Suppose we take one equivalent martingale measure $\mathbb{Q}^*$. Then, in terms of this "reference measure", we can give precise definitions for derivative securities and trading strategies given in the previous subsection.

A *derivative security* is a $\mathcal{F}_T$-measurable random variable $H(T)$ such that $\mathbb{E}^* \left(|H(T)| \right) < \infty$, where $\mathbb{E}^*$ denotes expectation under the measure $\mathbb{Q}^*$. Hence, derivative securities are those securities for which the expectation of the payoff is well-defined.

A *trading strategy* is a predictable N-dimensional stochastic process $\left(\delta_1(t,\omega), \ldots, \delta_N(t,\omega) \right)$ such that the stochastic integrals

$$\int_0^t \delta_n(s) \, dZ_n'(s) \tag{2.5}$$

are martingales under $\mathbb{Q}^*$. For self-financing strategies this implies that the value $V'(\delta, t)$ in terms of the relative prices Z' is a $\mathbb{Q}^*$-martingale.

The condition on trading strategies is a rather technical condition. It arises from the fact that for predictable processes in general, the Itô integrals that define the value processes $V'(\delta, t)$ of self-financing trading strategies yield only *local* martingales under $\mathbb{Q}^*$. For a local martingale

$$\sup_{t\in[0,T]} \left\{ \mathbb{E}^* \left(V'(\delta,t) \right) \right\} = \infty \qquad (2.6)$$

is possible, while for martingales

$$\sup_{t\in[0,T]} \left\{ \mathbb{E}^* \left(V'(\delta,t) \right) \right\} < \infty \qquad (2.7)$$

is always satisfied. This difference between local martingales and martingales allows for the existence of so-called *doubling strategies*, which are arbitrage opportunities. This was first pointed out by Harrison and Pliska (1981). Hence, an economy can only be arbitrage-free if the value processes of self-financing trading strategies are martingales.

Several restrictions can be imposed on the processes δ to ensure the martingale property of the value processes $V'(\delta,t)$. For example, one can show that the presence of wealth constraints or constraints like margin requirements ensures that the value processes are martingales. Because these constraints are actually present in security markets, it will be assumed throughout this book that this restriction holds.

Subject to the definitions given above, we have the following result:

Theorem (Unique Equivalent Martingale Measure) *A continuous economy is free of arbitrage opportunities and every derivative security is attainable if for every choice of numeraire there exists a unique equivalent martingale measure.*

We can paraphrase this extremely important result as follows. Given a choice of numeraire, we can find a unique probability measure such that the relative price processes are martingales. The martingale property is the mathematical reflection of the fact that in an arbitrage-free economy it is not possible to systematically outperform the market (hence the relative prices) by trading in the marketed assets. With the self-financing trading strategies in the marketed assets, we still cannot outperform the market, hence these trading strategies better be martingales as well. By a nice mathematical result called the martingale representation theorem one can clinch the completeness of the economy since every random variable (=a payoff pattern) can be represented (=replicated) as a stochastic integral with respect to a martingale (=a trading strategy in the underlying assets Z). Hence, we see that the language of martingales and stochastic processes provides a very powerful tool for describing arbitrage-free economies and replicating trading strategies.

Note that for different choices of numeraire there exists different unique equivalent martingale measure. To make things worse, the definition of derivative securities changes also with a different choice of numeraire. It is therefore conceivable that a payoff pattern $H(T)$ which can be replicated for one choice of numeraire, cannot be replicated for another choice of numeraire. Much work remains to be done in this area, for example in establishing which set of payoff patterns can be replicated for all choices of numeraire. However, for the derivatives we analyse in this book, we never encounter such a situation, and we will implicitly make the assumption that the payoff patterns we analyse can be replicated with any choice of numeraire.

From the result given above follows immediately that for a given numeraire M with unique equivalent martingale measure $\mathbb{Q}^M$, the value of a self-financing trading strategy $V'(\delta, t) = V(\delta, t)/M(t)$ is a $\mathbb{Q}^M$-martingale. Hence, for a replicating strategy δ_H that replicates the derivative security $H(T)$ we obtain

$$\mathbb{E}^M \left(\frac{H(T)}{M(T)} \,\Big|\, \mathcal{F}_t \right) = \mathbb{E}^M \left(\frac{V(\delta_H, T)}{M(T)} \,\Big|\, \mathcal{F}_t \right) = \frac{V(\delta_H, t)}{M(t)}, \qquad (2.8)$$

where the last equality follows from the definition of a martingale. Combining the first and last expression yields

$$V(\delta_H, t) = M(t)\mathbb{E}^M \left(\frac{H(T)}{M(T)} \,\Big|\, \mathcal{F}_t \right). \qquad (2.9)$$

This formula can be used to determine the value at time $t < T$ for any derivative security $H(T)$.

The theorem of the Unique Equivalent Martingale Measure was first proved by Harrison and Kreps (1979). In their paper they used the value of a riskless money-market account as the numeraire. Later it was recognised that the choice of numeraire is arbitrary. However, for this historic reason, the unique equivalent martingale measure obtained by taking the value of a money-market account as a numeraire is called "the" equivalent martingale measure, which is a very unfortunate name. In this book we will stick to this convention, because it is so widely used.

To illustrate the concepts developed here, we will apply them to the well known Black-Scholes (1973) framework. However, before we do so, we show several results we will be using extensively throughout the book for explicit calculations.

2.1.3 Change of Numeraire Theorem

Equation (2.9) shows how to calculate the value $V(t)$ of a derivative security. The value calculated must, of course, be independent of the choice of numeraire.

Consider two numeraires N and M with the martingale measures $\mathbb{Q}^N$ and $\mathbb{Q}^M$. Combining the result of (2.9) applied to both numeraires yields

$$N(t)\mathbb{E}^N \left(\frac{H(T)}{N(T)} \,\Big|\, \mathcal{F}_t \right) = M(t)\mathbb{E}^M \left(\frac{H(T)}{M(T)} \,\Big|\, \mathcal{F}_t \right). \qquad (2.10)$$

This expression can be rewritten as

$$\mathbb{E}^N\left(G(T)\,\big|\,\mathcal{F}_t\right) = \mathbb{E}^M\left(G(T)\frac{N(T)/N(t)}{M(T)/M(t)}\,\big|\,\mathcal{F}_t\right), \qquad (2.11)$$

where $G(T) = H(T)/N(T)$. Since, H, N and M are general, this result holds for all random variables G and all numeraires N and M.

We have now derived a way to express the expectation of $G(T)$ under the measure $\mathbb{Q}^N$ in terms of an expectation under the measure $\mathbb{Q}^M$. The expectation of G under $\mathbb{Q}^N$ is equal to the expectation of G times the random variable $\frac{N(T)/N(t)}{M(T)/M(t)}$ under the measure $\mathbb{Q}^M$. This random variable is known as the *Radon-Nikodym derivative* and is denoted by $d\mathbb{Q}^N/d\mathbb{Q}^M$.

The result we have just derived can be stated as follows:

Theorem (Change of Numeraire) *Let $\mathbb{Q}^N$ be the equivalent martingale measure with respect to the numeraire $N(t)$. Let $\mathbb{Q}^M$ be the equivalent martingale measure with respect to the numeraire $M(t)$. The Radon-Nikodym derivative that changes the equivalent martingale measure $\mathbb{Q}^M$ into $\mathbb{Q}^N$ is given by*

$$\frac{d\mathbb{Q}^N}{d\mathbb{Q}^M} = \frac{N(T)/N(t)}{M(T)/M(t)}. \qquad (2.12)$$

Proof. This result was first proven by Geman et al. (1995). The proof goes along the same lines as described above. □

2.1.4 Girsanov's Theorem and Itô's Lemma

The next two results will be stated without proof. For an introduction to the topics of Brownian motion and stochastic calculus, we refer to Øksendal (1998) or Karatzas and Shreve (1991).

A key result which can be used to explicitly determine equivalent martingale measures is *Girsanov's Theorem*. This theorem provides us with a tool to determine the effect of a change of measure on a stochastic process.

Theorem (Girsanov) *For any stochastic process $\kappa(t)$ such that*

$$\int_0^t \kappa(s)^2\,ds < \infty, \qquad (2.13)$$

with probability one, consider the Radon-Nikodym derivative $\frac{d\mathbb{Q}^*}{d\mathbb{Q}} = \rho(t)$ given by

$$\rho(t) = \exp\left\{\int_0^t \kappa(s)\,dW(s) - \tfrac{1}{2}\int_0^t \kappa(s)^2\,ds\right\}, \qquad (2.14)$$

where W is a Brownian motion under the measure $\mathbb{Q}$. Under the measure $\mathbb{Q}^*$ the process

$$W^*(t) = W(t) - \int_0^t \kappa(s)\,ds \qquad (2.15)$$

is also a Brownian motion.

The last equation in Girsanov's Theorem can be rewritten as

$$dW = dW^* + \kappa(t)\, dt \qquad (2.16)$$

which is a result we will often use.

Another key result from stochastic calculus is known as *Itô's Lemma*. Given a stochastic process x described by a stochastic differential equation, Itô's Lemma allows us to describe the behaviour of stochastic processes derived as functions $f(t, x)$ of the process x.

Lemma (Itô) *Suppose we have a stochastic process x given by the stochastic differential equation $dx = \mu(t, \omega)\, dt + \sigma(t, \omega)\, dW$ and a function $f(t, x)$ of the process x, then f satisfies*

$$df = \left(\frac{\partial f(t, x)}{\partial t} + \mu(t, \omega) \frac{\partial f(t, x)}{\partial x} + \tfrac{1}{2}\sigma(t, \omega)^2 \frac{\partial^2 f(t, x)}{\partial x^2} \right) dt$$
$$+ \sigma(t, \omega) \frac{\partial f(t, x)}{\partial x}\, dW, \quad (2.17)$$

provided that f is sufficiently differentiable.

2.1.5 Application: Black-Scholes Model

Let us now consider the Black and Scholes (1973) option pricing model. Using this familiar setting enables us to illustrate the concepts developed. In the Black-Scholes economy there are two marketed assets: B which is the value of a riskless money-market account with $B(0) = 1$ and a stock S. The prices of the assets are described by the following stochastic differential equations

$$dB = rB\, dt$$
$$dS = \mu S\, dt + \sigma S\, dW. \qquad (2.18)$$

The money-market account is assumed to earn a constant interest rate r, and the stock price is assumed to follow a geometric Brownian motion with constant drift μ and constant volatility σ.

The value of the money-market account is strictly positive and can serve as a numeraire. Hence, we obtain the relative price $S'(t) = S(t)/B(t)$. From Itô's Lemma we obtain that the relative price process follows

$$dS' = (\mu - r)S'\, dt + \sigma S'\, dW. \qquad (2.19)$$

To identify equivalent martingale measures we can apply Girsanov's Theorem. For $\kappa(t) \equiv -(\mu - r)/\sigma$ we obtain the new measure $\mathbb{Q}^B$ where the process S' follows

$$dS' = (\mu - r)S' \, dt + \sigma S' \left(dW^B - \tfrac{\mu - r}{\sigma} \, dt\right)$$
$$= \sigma S' \, dW^B \tag{2.20}$$

which is a martingale. For $\sigma \neq 0$ this is the only measure which turns the relative prices into martingales, and the measure $\mathbb{Q}^B$ is unique. Therefore, the Black-Scholes economy is arbitrage-free and complete for $\sigma \neq 0$.

Under the measure $\mathbb{Q}^B$, the original price process S follows the process

$$dS = \mu S \, dt + \sigma S \left(dW^B - \tfrac{\mu - r}{\sigma} \, dt\right)$$
$$= rS \, dt + \sigma S \, dW^B. \tag{2.21}$$

We see that under the equivalent martingale measure the drift μ of the process S is replaced by the interest rate r. The solution to this stochastic differential equation can be expressed as

$$S(t) = S(0) \exp\{(r - \tfrac{1}{2}\sigma^2)t + \sigma W^B(t)\}, \tag{2.22}$$

where $W^B(t)$ is the value of the Brownian motion at time t under the equivalent martingale measure. The random variable $W^B(t)$ has a normal distribution with mean 0 and variance t.

A European call option with strike K has at the exercise time T a payoff of $\mathbf{C}(T) = \max\{S(T) - K, 0\}$. From (2.9) follows that the price of the option $\mathbf{C}(0)$ at time 0 is given by $\mathbb{E}^B\left(\max\{S(T) - K, 0\}/B(T)\right)$. To evaluate this expectation, we use the explicit solution of $S(T)$ under the equivalent martingale measure given in (2.22) and we get

$$\mathbb{E}^B\left(\max\{S(T) - K, 0\}/B(T)\right) =$$
$$\int_{-\infty}^{\infty} e^{-rT} \max\{S(0)e^{(r - \frac{1}{2}\sigma^2)T + \sigma w} - K, 0\} \frac{e^{-\frac{1}{2}\frac{w^2}{T}}}{\sqrt{2\pi T}} \, dw. \tag{2.23}$$

A straightforward calculation will confirm that this integral can be expressed in terms of cumulative normal distribution functions $N(.)$ as follows

$$\mathbf{C}(0) = S(0)N(d) - e^{-rT}KN(d - \sigma\sqrt{T}) \tag{2.24}$$

with

$$d = \frac{\log\left(\frac{S(0)}{K}\right) + (r + \tfrac{1}{2}\sigma^2)T}{\sigma\sqrt{T}} \tag{2.25}$$

which is the celebrated Black-Scholes option pricing formula.

In the derivation given above, we used the value of a money-market account B as a numeraire. However, this choice is arbitrary. The stock price S is also strictly positive for all t and can also be used as a numeraire. If we choose S as a numeraire, we obtain from Itô's Lemma that the relative price $B' = B/S$ follows

$$dB' = (r - \mu + \sigma^2)B' \, dt - \sigma B' \, dW. \tag{2.26}$$

If we apply Girsanov's Theorem with $\kappa = (r - \mu)/\sigma + \sigma$, we obtain (for $\sigma \neq 0$) the unique equivalent martingale measure $\mathbb{Q}^S$ for which the relative price B' is a martingale. From (2.9) we obtain

$$
\begin{aligned}
\mathbf{C}(0) &= S(0)\mathbb{E}^S \left(\frac{\max\{S(T) - K, 0\}}{S(T)} \right) \\
&= S(0)\mathbb{E}^S \left(\max\{1 - K\tfrac{1}{S(T)}, 0\} \right).
\end{aligned}
\tag{2.27}
$$

Using Itô's Lemma and Girsanov's Theorem we obtain for the equivalent martingale measure $\mathbb{Q}^S$, that the process $1/S$ follows

$$
\begin{aligned}
d\frac{1}{S} &= (-\mu + \sigma^2)\frac{1}{S} dt - \sigma\frac{1}{S}\left(dW^S + (\tfrac{r-\mu}{\sigma} + \sigma)dt\right) \\
&= -r\frac{1}{S} dt - \sigma\frac{1}{S} dW^S,
\end{aligned}
\tag{2.28}
$$

where W^S is a Brownian motion under $\mathbb{Q}^S$. The explicit solution can be expressed as

$$
\frac{1}{S(t)} = \frac{1}{S(0)} \exp\{(-r - \tfrac{1}{2}\sigma^2)t - \sigma W^S(t)\}.
\tag{2.29}
$$

Using this explicit form, we can evaluate the expectation (2.27). It is left to the reader to verify that this also gives the Black-Scholes formula (2.24).

2.1.6 Application: Foreign-Exchange Options

The example for the Black-Scholes economy given above is a bit contrived. However a more fruitful application can be found when we consider foreign-exchange (F/X) options. The first valuation formula for F/X-options in a Black-Scholes setting was given by Garman and Kohlhagen (1983). This formula is nowadays widely used by F/X-option traders all over the world.

An interesting aspect of F/X-derivatives is that we can either calculate the value of a derivative in the domestic market or in the foreign market. If the economy is arbitrage-free, both values must be the same, otherwise an "international" arbitrage opportunity would arise.

Consider the following, very simple, international economy. In the domestic market D there is a money-market account B^D, which earns a riskless instantaneous interest rate r^D; in the foreign country F there is also a money-market account B^F with interest rate r^F. Furthermore, the exchange rate X follows a geometric Brownian motion. The three price processes can be summarised as

$$
\begin{cases}
dB^F = r^F B^F \, dt \\
dX = \mu X \, dt + \sigma X \, dW \\
dB^D = r^D B^D \, dt.
\end{cases}
\tag{2.30}
$$

From a domestic point of view, there are two marketed assets: the domestic money-market account B^D and the value of the foreign money-market account in domestic terms, given by $B^F X$. From Itô's Lemma we obtain that the process $(B^F X)$ follows

$$d(B^F X) = (r^F + \mu)(B^F X)\,dt + \sigma(B^F X)\,dW. \qquad (2.31)$$

The domestic money-market account can be used as a numeraire, and the relative price process $(B^F X)' = (B^F X)/B^D$ follows the process

$$d(B^F X)' = (r^F - r^D + \mu)(B^F X)'\,dt + \sigma(B^F X)'\,dW. \qquad (2.32)$$

An application of Girsanov's Theorem with $\kappa(t) \equiv -(r^F - r^D + \mu)/\sigma$ will yield the domestic unique equivalent martingale measure $\mathbb{Q}^D$ under which the relative price process $(B^F X)'$ is a martingale. Under the domestic measure $\mathbb{Q}^D$, the exchange rate process follows

$$dX = (r^D - r^F)X\,dt + \sigma X\,dW^D, \qquad (2.33)$$

which is the process used in the Garman-Kohlhagen formula.

We can also take the perspective of the foreign market. Here we also have two marketed assets: B^F and (B^D/X). Using B^F as a numeraire, we obtain the relative price process $(B^D/X)' = (B^D/X)/B^F$ which follows the process

$$d\left(\frac{B^D}{X}\right)' = (r^D - r^F - \mu + \sigma^2)\left(\frac{B^D}{X}\right)'\,dt - \sigma\left(\frac{B^D}{X}\right)'\,dW. \qquad (2.34)$$

If we apply Girsanov's Theorem with $\kappa(t) \equiv (r^D - r^F - \mu)/\sigma + \sigma$, we obtain the foreign unique equivalent martingale measure $\mathbb{Q}^F$. Under the foreign measure $\mathbb{Q}^F$, the foreign exchange rate $1/X$ follows the process

$$d\left(\frac{1}{X}\right) = (r^F - r^D)\left(\frac{1}{X}\right)\,dt - \sigma\left(\frac{1}{X}\right)\,dW^F. \qquad (2.35)$$

This process is exactly the right process for calculating the Garman-Kohlhagen formula in the foreign market. Hence, in this economy a trader in the domestic market and a trader in the foreign market will calculate exactly the same price for a F/X-option.

For more examples of calculating prices of derivatives under domestic and foreign martingale measures see Reiner (1992) or Hull (2000), where these measures are used repeatedly to calculate the value of so-called *quanto options*, which are options on foreign assets denominated in the domestic currency. We will give examples of interest rate quanto options (also known as *diff options*) in Chapter 11.

2.2 Numerical Methods

In this section we give a brief overview of a different methodology for valuing options. By exploiting the fact that for every financial instrument a replicating portfolio can be found and by using no-arbitrage arguments, a partial differential equation can be derived that describes the value of a financial instrument through time.

Given the fact that efficient numerical methods can be used to solve partial differential equations, it is often possible to obtain an accurate approximation of the price of a financial instrument from a partial differential equation in cases where the explicit evaluation of the expectation under the equivalent martingale measure is very difficult. One of the best known examples is probably the pricing of American-style options.

The academic finance literature devotes relatively little attention to the subject of partial differential equations, because it is considered to be more an engineering than an academic problem. However, we believe that no option pricing model can be implemented successfully without a thorough understanding of this subject. The interested reader is referred to the books by Wilmott, Dewynne and Howison (1993) and Wilmott (1998) which are largely devoted to the use and solution methods of partial differential equations applied to option pricing theory.

First, we show how in the Black-Scholes economy the partial differential equation can be derived. This is in fact the method employed in the original paper of Black and Scholes (1973). Then we show how prices of options can calculated using Monte Carlo simulation and numerical integration.

2.2.1 Derivation of Black-Scholes Partial Differential Equation

As in Section 2.1.4 we assume a continuous-time economy with two marketed assets B and S that follow the processes given in (2.18). We furthermore assume that the value V of a financial instrument is completely determined at every instant t by the asset price $S(t)$. Hence, the value is a function $V(t, S)$. By making this assumption we restrict ourselves to financial instruments whose value does not depend on the history of asset prices until time t. Applying Itô's Lemma gives the following stochastic differential equation for V

$$dV = \left(V_t + \mu S V_S + \tfrac{1}{2}\sigma^2 S^2 V_{SS}\right)dt + \sigma S V_S \, dW, \qquad (2.36)$$

where subscripts on the function V denote partial derivatives.

On the other hand, we can replicate the value $V(t, S)$ with a self-financing trading strategy δ such that $V(t, S) = \delta_S(t)S(t) + \delta_B(t)B(t)$. Writing the definition of a self-financing trading strategy (2.4) in differential form, we obtain $dV = \delta_S \, dS + \delta_B \, dB$. We can simplify this expression for the Black-Scholes economy using (2.18) and we obtain

$$dV = \big(r(V - \delta_S S) + \mu S \delta_S\big)dt + \sigma S \delta_S \, dW. \tag{2.37}$$

Equating both expressions for dV from (2.36) and (2.37) yields

$$\big(V_t - r(V - \delta_S S) + \mu S(V_S - \delta_S) + \tfrac{1}{2}\sigma^2 S^2 V_{SS}\big)dt + \sigma S(V_S - \delta_S)dW = 0. \tag{2.38}$$

If we choose $\delta_S \equiv V_S$, we see that the dW term disappears for all t. Simplifying for this choice of δ_S leads to

$$V_t + rSV_S + \tfrac{1}{2}\sigma^2 S^2 V_{SS} - rV = 0, \tag{2.39}$$

which is the Black-Scholes partial differential equation.

Prices of financial instruments can be calculated by solving the partial differential equation with respect to a boundary condition that describes the payoff of the instrument at time T. For example, the price of a European option is given by solving (2.39) with respect to the boundary condition describing the payoff at time T, namely $V(T, S) = \max\{S(T) - K, 0\}$.

2.2.2 Feynman-Kac Formula

A method of solving partial differential equations like (2.39) is to use the *Feynman-Kac formula*.

Theorem (Feynman-Kac) *The partial differential equation*

$$V_t + \mu(t, x)V_x + \tfrac{1}{2}\sigma(t, x)^2 V_{xx} - r(t, x)V = 0$$

with boundary condition $H(T, x)$ has solution

$$V(t, x) = \mathbb{E}\left(e^{-\int_t^T r(s, X)\, ds} H(T, X)\right),$$

where the expectation is taken with respect to the process X defined by

$$dX = \mu(t, X)\, dt + \sigma(t, X)\, dW.$$

Proof. A proof of the Feynman-Kac formula can be found in Øksendal (1998). Proofs of generalised versions of the Feynman-Kac formula can be found in Rogers and Williams (1994). □

The Black-Scholes partial differential equation (2.39) can be solved with the Feynman-Kac formula. If we substitute $x = S$, $\mu(t, x) = rS$, $\sigma(t, x) = \sigma S$ and $r(t, x) = r$, we can express the solution with respect to a final payoff function $H(T, x)$ as

$$V(t, S) = \mathbb{E}\left(e^{-r(T-t)} H(T, S^*)\right), \tag{2.40}$$

where $\mathbb{E}$ denotes the expectation with respect to the process S^*

$$dS^* = rS^* \, dt + \sigma S^* \, dW. \tag{2.41}$$

We see that the expectation operator $\mathbb{E}$ with respect to the process S^* is exactly the same as the expectation of the discounted payoff under the equivalent martingale measure $\mathbb{Q}^B$ if we choose B as numeraire. Hence, calculating the Black-Scholes formula using the partial differential equation leads to exactly the same result (2.24) which was obtained in Section 2.1.5.

2.2.3 Numerical Solution of PDE's

Although we have put a lot of emphasis on analytical formulæ up until now, it is often true that prices for options cannot be calculated analytically. All numerical methods seek to find a solution for the martingale pricing relationship (2.9). For a practical implementation of a model for pricing interest rate derivatives (for example, in a trading environment) it is, of course, very important to use a numerical implementation of the model that can calculate accurate answers sufficiently fast.

One method is the numerical solution of the pricing partial differential equation. The solution of the partial differential equation $V(t, S)$ is approximated by values $V_{i,j}$ on a grid with points (t_i, S_j). On the grid the partial derivatives of the solution can be calculated via finite difference approximations. The partial differential equation imposes a relation between the "spatial" derivatives (S-derivatives) and the time-derivative. This relation is then used to propagate the solution from the boundary condition at time T backward to the initial time 0.

Different choices for the grid spacing, and different differencing schemes lead to different algorithms for solving partial differential equations. For detailed derivations of several algorithms, and a discussion of stability and convergence for different algorithms, see Wilmott, Dewynne and Howison (1993). In Chapters 5 and 6 we provide a derivation of explicit finite difference algorithms for spot interest rate models.

2.2.4 Monte Carlo Simulation

Suppose we are given for the process x a general stochastic differential equation

$$dx(t) = \mu(t) \, dt + \sigma(t) \, dW(t), \tag{2.42}$$

where μ and σ are allowed to be stochastic (i.e. allowed to depend on the history of W). This stochastic differential equation is only a notational shorthand for the stochastic integral equation

$$x(T) = x(t) + \int_t^T \mu(s) \, ds + \int_t^T \sigma(s) \, dW(s) \tag{2.43}$$

for the time interval $[t, T]$. For a short time interval $[t, t + \Delta t]$, we can *approximate* the integrals by

$$x(t + \Delta t) = x(t) + \mu(t)\Delta t + \sigma(t)\big(W(t + \Delta t) - W(t)\big). \tag{2.44}$$

From the definition of Brownian motion we know that the increments of the Brownian motion $W(t + \Delta t) - W(t)$ are independent random variables which have a normal distribution with mean 0 and variance Δt.

A path for the Brownian motion W can now be constructed as follows. Given the initial value $W(0) = 0$ and a time-step Δt, we can construct the approximation at times $t_i = i\,\Delta t$ for $i = 1, 2, \ldots$ as

$$W(t_{i+1}) = W(t_i) + \sqrt{\Delta t}\,\epsilon_i, \tag{2.45}$$

where the ϵ_i are independent standard normal random variables.

Let us consider a simple example that can be recreated easily in a spreadsheet. Suppose we have the stochastic differential equation $dx = \sigma x\, dW$ where σ is a constant. For this simple example we know that the solution can be expressed as

$$x(t + \Delta t) = x(t)\exp\big\{-\tfrac{1}{2}\sigma^2\Delta t + \sigma\big(W(t + \Delta t) - W(t)\big)\big\}. \tag{2.46}$$

The approximate solution is constructed as

$$x(t + \Delta t) = x(t) + \sigma x(t)\big(W(t + \Delta t) - W(t)\big). \tag{2.47}$$

We see that the approximate solution is not exact for this example. To assess the accuracy of the approximation we have compiled Table 2.1.

Table 2.1. Approximation of stochastic process

$dx = \sigma x\, dW$ with $x_0 = 5\%$, $\sigma = 0.10$

t	$\Delta t = 1$		$\Delta t = 1/12$	
	exact	appr	exact	appr
0	5.00%	5.00%	5.00%	5.00%
1	5.07%	5.09%	5.71%	5.72%
2	4.67%	4.70%	6.18%	6.17%
3	4.12%	4.14%	6.44%	6.44%
4	4.26%	4.29%	6.26%	6.27%
5	4.53%	4.58%	6.84%	6.84%
6	4.32%	4.38%	6.20%	6.19%
7	3.96%	4.03%	6.81%	6.80%
8	4.20%	4.28%	7.39%	7.37%
9	4.26%	4.37%	7.16%	7.13%
10	4.19%	4.31%	7.57%	7.53%

We have simulated the process x for a period of 10 years both with annual time-steps ($\Delta t = 1$) and with monthly time-steps ($\Delta t = 1/12$). For the monthly simulation we have only reported the results on the year points (otherwise we would have 120 entries in the table). We see that the simulation with annual time-steps is already reasonably accurate with the difference to the exact solution only a few tenths of a percent. The simulation with monthly steps is very accurate, with errors of only a few basispoints.

Suppose we draw M paths in this way for the process x, and for each path we calculate the payoff V_j' for $j = 1, \ldots, M$. Because the M drawings are independent and from the same (but usually unknown) probability distribution with mean $\mathbb{E}(V')$ and variance $\mathrm{Var}(V')$ we know from the central limit theorem that the probability distribution of the random variable

$$V^* = \frac{1}{M} \sum_{j=1}^{M} V_j' \qquad (2.48)$$

converges for large M to a normal distribution with mean $\mathbb{E}(V')$ and variance $1/M \mathrm{Var}(V')$. We see that for increasing M, the variance of V^* decreases. Hence, V^* becomes a more accurate estimate of $\mathbb{E}(V')$ for increasing M.

When presenting numerical results, we report for a given M the random variable V^* and the standard deviation of V^* which is called the *standard error* of the Monte Carlo simulation. The standard error is calculated as

$$\mathrm{stderr}(V^*) = \sqrt{\frac{\sum_{j=1}^{M} V_j'^2 - M(V^*)^2}{M(M-1)}}. \qquad (2.49)$$

In this section we have only treated the very basics of Monte Carlo simulation. The interested reader is referred to Rebonato (1998, Chapter 10) and Press et al. (1992, Chapter 7). These references treat so-called variance reduction techniques which for a given number of paths M seek to reduce the standard error of the simulation.

In Chapter 8 we give examples how to calculate prices within Market Models using Monte Carlo simulation.

2.2.5 Numerical Integration

One can also try to solve the expectation (2.9) by direct numerical integration of the payoff against the probability distribution of the underlying assets. This methodology can be implemented in various ways by choosing different algorithms for the numerical integration. For an introduction to numerical integration algorithms, we refer to Press et al. (1992, Chapter 4).

In Chapter 9 we show how to implement this method to calculate prices in the Markov-Functional model.

Spot and Forward Rate Models

3. Spot and Forward Rate Models

The first part of this book is devoted to *spot and forward rate models*. These types of models take instantaneous interest rates as the basis for modelling the term-structure of interest rates. A spot instantaneous interest rate is the interest one earns on a riskless investment over an infinitesimal time-period dt. Once we have a model for the evolution of the spot instantaneous interest rates, all other interest rates can be derived by integrating over the spot interest rates.

In this chapter we explain the theory of the pricing of interest rate derivatives, and we point out the implications of the fact that we cannot trade in the spot interest rate. Section 3.1 is devoted to the valuation of interest rate derivatives using a partial differential equation approach, using the methodology of Vasicek (1977). In Section 3.2 we explain how interest rate derivatives can be priced via expectations under the equivalent martingale measure, using the methodology of Heath, Jarrow and Morton (1992). In the final section we discuss similarities between the two methodologies.

3.1 Vasicek Methodology

In this section we show how interest rate derivatives can be valued using partial differential equations. The derivation of the partial differential equation is based on Vasicek (1977).

3.1.1 Spot Interest Rate

We will restrict our attention to one-factor models, which describe the evolution of the spot interest rate with one source of uncertainty. Again, the generalisation to multi-factor models is quite straightforward, but makes the notation more complicated.

We can write down the following general stochastic differential equation for the spot interest rate r

$$dr = \mu(t, r)\, dt + \sigma(t, r)\, dW. \tag{3.1}$$

In this general form the functions $\mu(t,r)$ and $\sigma(t,r)$ are left unspecified. Different choices for the functions μ and σ give rise to different models.

Given this stochastic process for the spot interest rate r we can proceed to derive a partial differential equation by constructing a locally riskless portfolio.

3.1.2 Partial Differential Equation

In analogy to Chapter 2 we want to consider the value V of financial instruments whose value is determined at time t by the value of the spot interest rate $r(t)$. The value of a financial instrument V is a function $V(t,r)$. From Itô's Lemma we obtain

$$dV = M(t,r)\, dt + \Sigma(t,r)\, dW, \tag{3.2}$$

with

$$
\begin{aligned}
M(t,r) &= V_t + \mu(t,r)V_r + \tfrac{1}{2}\sigma(t,r)^2 V_{rr} \\
\Sigma(t,r) &= \sigma(t,r)V_r
\end{aligned}
\tag{3.3}
$$

Let us now attempt to construct a locally riskless portfolio Π. We would like to take a position in the instrument V with a short position in the spot interest rate r. Unfortunately, the spot interest rate r is not a traded asset, so this is impossible. The best thing we can do, is to hedge a derivative V_1 with another interest rate derivative V_2 to obtain the portfolio

$$\Pi = V_1(t,r) - \Delta V_2(t,r), \tag{3.4}$$

where V_1 and V_2 follow processes similar to (3.2), with coefficients M_1, Σ_1 and M_2, Σ_2, respectively.

The portfolio Π is a linear combination of the stochastic processes V_1 and V_2. Hence, we obtain

$$d\Pi = \big(M_1(t,r) - \Delta M_2(t,r)\big)dt + \big(\Sigma_1(t,r) - \Delta\Sigma_2(t,r)\big)dW. \tag{3.5}$$

For the choice $\Delta = \Sigma_1/\Sigma_2$ the dW disappears and the portfolio becomes locally riskless for the time interval dt. To avoid arbitrage opportunities, the portfolio must earn in this case the locally riskless return, which is the spot interest rate r. Hence, for this choice of Δ we obtain for the time period dt

$$d\Pi = r\Pi\, dt$$

$$\Downarrow$$

$$\left(M_1(t,r) - \frac{\Sigma_1(t,r)}{\Sigma_2(t,r)}M_2(t,r)\right)dt = r\left(V_1(t,r) - \frac{\Sigma_1(t,r)}{\Sigma_2(t,r)}V_2(t,r)\right)dt. \tag{3.6}$$

Dividing by dt and rearranging terms leads to

$$\frac{M_1(t,r) - rV_1(t,r)}{\Sigma_1(t,r)} = \frac{M_2(t,r) - rV_2(t,r)}{\Sigma_2(t,r)}. \tag{3.7}$$

This equality must hold for any pair of derivatives V_1 and V_2, which is only possible if the value of the ratio $(M - rV)/\Sigma$ is a function of t and r only. Let $\lambda(t, r)$ denote the common value of the ratio, which is known as the *market price of risk*. Hence, any interest rate derivative V must satisfy

$$\frac{M(t, r) - rV(t, r)}{\Sigma(t, r)} = \lambda(t, r), \tag{3.8}$$

where M and Σ are defined as in (3.3). Substituting these definitions into (3.8) and rearranging terms yields

$$V_t + \big(\mu(t, r) - \lambda(t, r)\sigma(t, r)\big) V_r + \tfrac{1}{2}\sigma(t, r)^2 V_{rr} - rV = 0 \tag{3.9}$$

which is the partial differential equation that describes the prices of securities in one-factor yield-curve models.

3.1.3 Calculating Prices

Armed with the partial differential equation (3.9), the price of an interest rate derivative V with a payoff $V(T, r)$ at time T can be calculated with the help of the Feynman-Kac formula. The price $V(t, r)$ is the solution to the partial differential equation subject to the boundary condition $V(T, r)$ and can be expressed as

$$V(t, r) = \mathbb{E}\left(e^{-\int_t^T r^*(s)\, ds} V(T, r^*) \right), \tag{3.10}$$

where the expectation is taken with respect to the process r^*

$$dr^* = \theta(t, r^*)\, dt + \sigma(t, r^*)\, dW, \tag{3.11}$$

where

$$\theta(t, r^*) = \mu(t, r^*) - \lambda(t, r^*)\sigma(t, r^*). \tag{3.12}$$

The process r^* used to calculate the expectation is different from the process r defined in (3.1). The dW term of both processes is the same, but the drift term of r^* is corrected by a factor involving the market price of risk $\lambda(t, r)$.

Before prices of derivatives can be calculated, the model has to be fitted to the initial term-structure of interest rates. The initial term-structure of interest rates is described by the prices of all discount bonds $D(0, T)$ at time $t = 0$. The payoff of a discount bond at maturity T is 1 in all states of the world. Hence, using the Feynman-Kac formula (3.10) we can express the discount bond prices in the functions θ and σ. Given a choice for σ, we can solve for θ from the initial term-structure of interest rates. When we have determined θ, we have actually estimated the drift μ and the market price of risk λ simultaneously from the initial term-structure of interest rates. The valuation formula (3.10) is then no longer (explicitly) dependent on the market price of risk.

3.1.4 Example: Ho-Lee Model

To illustrate the procedure outlined above we provide a simple example. If we assume that $\sigma(t, r)$ is a constant σ and that $\mu(t, r)$ is a function $\mu(t)$ of time only, we obtain the continuous time limit of the Ho and Lee (1986) model. For these choices of μ and σ the partial differential equation (3.9) reduces to

$$V_t + \left(\mu(t) - \lambda(t, r)\sigma\right)V_r + \tfrac{1}{2}\sigma^2 V_{rr} - rV = 0. \tag{3.13}$$

If we make the additional assumption that the market price of risk is a function $\lambda(t)$ of time only, the drift term is a function of time only which can be denoted by $\theta(t)$.

Using the Feynman-Kac formula, the prices of interest rate derivatives can be expressed as

$$V(t, r) = \mathbb{E}\left(e^{-\int_t^T r^*(s)\, ds} V(T, r^*)\right), \tag{3.14}$$

where the expectation is taken with respect to the process r^*

$$dr^* = \theta(t)\, dt + \sigma\, dW. \tag{3.15}$$

To fit this model to the initial term-structure of interest rates, we have to calculate the prices of discount bonds in terms of $\theta(t)$. The payoff of a discount bond at maturity is equal to 1, hence we have $V(T, r^*) \equiv 1$ and the price D of a discount bond is given by

$$D(0, T) = \mathbb{E}\left(e^{-y(T)}\right), \tag{3.16}$$

where the random variable y is defined as

$$y(t) = \int_0^t r^*(s)\, ds. \tag{3.17}$$

Substituting the solution of the stochastic differential equation (3.15) for r^* into the definition of $y(t)$ yields

$$y(t) = \int_0^t r_0\, ds + \int_0^t \int_0^s \theta(u)\, du\, ds + \int_0^t \int_0^s \sigma\, dW(u)\, ds. \tag{3.18}$$

By interchanging the order of integration and simplifying we obtain

$$y(t) = r_0 t + \int_0^t \theta(u)(t - u)\, du + \int_0^t \sigma(t - u)\, dW(u). \tag{3.19}$$

(For a proof of Fubini's Theorem for stochastic integrals, see the Appendix of Heath, Jarrow and Morton (1992).) Hence, the process $y(t)$ has a normal distribution with mean

$$m(t) = r_0 t + \int_0^t \theta(s)(t - s)\, ds \tag{3.20}$$

and variance

$$v(t) = \int_0^t \sigma^2 (t - s)^2\, ds = \tfrac{1}{3}\sigma^2 t^3. \tag{3.21}$$

From this follows that the expectation in the Feynman-Kac formula (3.16) can be evaluated as

$$D(0, T) = \exp\{-m(T) + \tfrac{1}{2}v(T)\}$$
$$= \exp\left\{-r_0 T - \int_0^T \theta(s)(T - s)\, ds + \tfrac{1}{6}\sigma^2 T^3\right\}. \tag{3.22}$$

The Ho-Lee model can be fitted to the initial term-structure of interest rates by solving for $\theta(t)$. Taking logarithms and differentiating twice with respect to T yields

$$\theta(T) = -\frac{\partial^2}{\partial T^2} \log D(0, T) + \sigma^2 T. \tag{3.23}$$

3.2 Heath-Jarrow-Morton Methodology

In this section we show how prices of interest rate derivatives can be calculated using equivalent martingale measures, which is the methodology of Heath, Jarrow and Morton (1992). We will only derive the equivalent martingale measure for one-factor interest rate models, this will allow us to explain the essence of the Heath-Jarrow-Morton (HJM) methodology.

3.2.1 Forward Rates

The marketed assets which can be traded are discount bonds with different maturities. The price of a discount bond at time t with maturity T is denoted by $D(t, T)$. In their setup HJM choose not to model discount bond prices directly, but to model the prices of *forward rates* $f(t, T)$. The forward rate is defined as

$$f(t, T) = \frac{-\partial \log D(t, T)}{\partial T}; \tag{3.24}$$

it is the instantaneous interest rate one can contract for at time t to invest in the money-market account at time T. It is easy to see that the spot interest rate $r(t)$ is equal to $f(t, t)$. HJM assume that the forward rates satisfy the following equation

$$f(t, T) - f(0, T) = \int_0^t \alpha(s, T, \omega)\, ds + \int_0^t \sigma(s, T, \omega)\, dW(s), \tag{3.25}$$

where ω denotes the state of the world. Equation (3.25) is the integral form of the stochastic differential equation

$$df(t, T) = \alpha(t, T, \omega)\, dt + \sigma(t, T, \omega)\, dW. \tag{3.26}$$

However, the integral form (3.25) of the equation is more precise. The stochastic process for the forward rates defined above is very general. The functions α and σ are allowed to depend on the maturity T of the forward rate and are allowed to depend on the state of the world ω.

The spot interest rate $r(t)$ is equal to $f(t, t)$; hence we get from (3.25)

$$r(t) = f(0, t) + \int_0^t \alpha(s, t, \omega)\, ds + \int_0^t \sigma(s, t, \omega)\, dW(s). \tag{3.27}$$

This stochastic process for the spot rate r is much more general than the process (3.1) proposed in the previous section. For the appropriate choices for α and σ it is (in principle) possible to reduce (3.27) to the form (3.1).

Using (3.24) we can express the discount bond prices in terms of the forward rates as

$$\log D(t, T) = - \int_t^T f(t, s)\, ds. \tag{3.28}$$

Substituting (3.25) into this equation and by interchanging the order of integration and simplifying, HJM obtain the following process for the discount bond prices (suppressing the notational dependence on ω)

$$dD(t, T) = b(t, T)D(t, T)\, dt + a(t, T)D(t, T)\, dW, \tag{3.29}$$

where

$$\begin{aligned} a(t, T, \omega) &= - \int_t^T \sigma(t, s, \omega)\, ds \\ b(t, T, \omega) &= r(t) - \int_t^T \alpha(t, s, \omega)\, ds + \tfrac{1}{2}a(t, T, \omega)^2. \end{aligned} \tag{3.30}$$

3.2.2 Equivalent Martingale Measure

Having specified the stochastic process followed by the discount bonds $D(t, T)$ which are the marketed assets, we want to establish the existence of an equivalent martingale measure to ensure that no arbitrage opportunities can exist in the economy.

Suppose we keep reinvesting money in the money-market account. Every instant dt the money-market account earns the riskless spot interest rate and the value $B(t)$ of the money-market account is given by $dB = rB\, dt$. If we solve this ordinary differential equation we get

$$B(t) = \exp\left\{ \int_0^t r(s)\, ds \right\}. \tag{3.31}$$

As in the Black-Scholes economy of Chapter 2, the value of the money-market account is strictly positive and can be used as a numeraire. Hence, in the HJM economy we obtain the relative prices $D'(t,T) = D(t,T)/B(t)$. Itô's Lemma yields

$$dD'(t,T) = \big(b(t,T) - r(t)\big)D'(t,T)\,dt + a(t,T)D'(t,T)\,dW. \tag{3.32}$$

The HJM economy will be arbitrage-free if we can find a unique equivalent probability measure such that the relative prices D' of the discount bonds become martingales.

Suppose we consider the discount bond with maturity T_1. If we apply Girsanov's Theorem with $\kappa(t,T_1) = -\big(b(t,T_1) - r(t)\big)/a(t,T_1)$, we obtain under the new measure $\mathbb{Q}^{T_1}$ that the process $D'(t,T_1)$ is a martingale. This change of measure depends on the maturity of the discount bond T_1 and will only make this particular discount bond a martingale.

However, we want to find an equivalent martingale measure that changes all marketed assets, that is all discount bonds, to martingales. This is only possible if the ratio $\big(b(t,T,\omega) - r(t)\big)/a(t,T,\omega)$ is independent of T. Let $\lambda(t,\omega)$ denote the common value of this ratio, if we then apply Girsanov's Theorem with $\kappa(t,\omega) = -\lambda(t,\omega)$ we get that all discount bonds $D'(t,T)$ are martingales under the equivalent martingale measure $\mathbb{Q}^*$.

Since the prices of all discount bonds are dependent on the spot interest rate r, the drift term $b(t,T,\omega)$ cannot be specified arbitrarily. A unique equivalent martingale measure can only be found if the drift term is of the form

$$b(t,T,\omega) - r(t) = \lambda(t,\omega)a(t,T,\omega). \tag{3.33}$$

Substituting the definitions for a and b given in (3.30), and differentiating with respect to T we find that the drift terms of the forward rate processes $\alpha(t,T,\omega)$ are restricted to

$$\alpha(t,T,\omega) = \sigma(t,T,\omega)\left(\int_t^T \sigma(t,s,\omega)\,ds + \lambda(t,\omega)\right). \tag{3.34}$$

3.2.3 Calculating Prices

Now that we have determined under which conditions an equivalent martingale measure exists in the HJM model, we can calculate the prices of interest rate derivatives. In Chapter 2 we derived the result that under the equivalent martingale measure the relative prices $V(t,r)/B(t)$ are martingales. In the HJM economy we get that the price of a financial instrument with a payoff $H(T,r)$ at time T is given by

$$V(t,r) = \mathbb{E}^*\left(e^{-\int_t^T r(s)\,ds} H(T,r) \,\Big|\, \mathcal{F}_t\right) \tag{3.35}$$

where the expectation $\mathbb{E}^*$ is taken with respect to the equivalent martingale measure $\mathbb{Q}^*$. From Girsanov's Theorem and using the restriction on α given in (3.34) we obtain that under the equivalent martingale measure the process r follows

$$r(t) = f(0,t) + \int_0^t \sigma(s,t,\omega) \int_s^t \sigma(s,u,\omega)\, du\, ds + \int_0^t \sigma(s,t,\omega)\, dW^*(s). \quad (3.36)$$

It is clear that for a given initial term-structure of interest rates and for a given choice of the function $\sigma(t,T,\omega)$ the spot rate process under the equivalent martingale measure is completely determined.

3.2.4 Example: Ho-Lee Model

To illustrate the HJM methodology, we turn again to the continuous-time Ho-Lee model. If we set the function $\sigma(t,T,\omega)$ to a constant σ, and if we make the assumption that the market price of risk is a function $\lambda(t)$ of time only, we obtain from (3.34) that the drift terms of the forward rates are restricted to

$$\alpha(t,T) = \sigma\big(\sigma(T-t) + \lambda(t)\big). \quad (3.37)$$

Hence, under the equivalent martingale measure, the spot interest rate follows the process

$$r(t) = f(0,t) + \tfrac{1}{2}\sigma^2 t^2 + \sigma W^*(t), \quad (3.38)$$

which can also be written in differential form as

$$dr = \left(-\frac{\partial^2}{\partial t^2} \log D(0,t) + \sigma^2 t\right) dt + \sigma\, dW^*, \quad (3.39)$$

where we have used the definition of the forward rates given in (3.24).

3.3 Equivalence of the Methodologies

If we compare the process r^* from the Feynman-Kac formula with the process r under the equivalent martingale measure we see that they look very different. This has two reasons. First, the process for the forward rates defined by HJM is much more general than the forward rates induced by the process (3.1). In general, HJM processes are path dependent, which means that the stochastic process at time t depends on the path followed by the forward rates up until time t. Second, the drift term of the process r^* depends implicitly on the forward rate, because we still have to fit it to the initial term-structure of interest rates. The drift term of the process r under the equivalent martingale measure depends explicitly on the initial term-structure of interest rates.

However, it is possible to establish a general equivalence between the two methodologies. For the Ho-Lee model we have shown that both methodologies lead to exactly the same process for the spot interest rate. In fact, one can prove the result that every forward rate model can be written as a spot rate model, but the proof goes beyond the scope of this book. For a derivation of this result we refer to Hunt and Kennedy (2000), Chapter 8.

4. Fundamental Solutions and the Forward-Risk-Adjusted Measure

Prices of interest rate derivatives can be calculated as the expected value of the discounted payoff, this was explained in the previous chapter. However, interest rates play a double role in interest rate models: they determine the amount of discounting, and they determine the payoff of the security. This implies that the discounting term and the payoff term are two correlated stochastic variables, which makes the evaluation of the expectation quite difficult.

As was shown independently by Jamshidian (1991) and Geman et al. (1995), one can use the T-maturity discount bond as a numeraire with its associated unique equivalent martingale measure. Under this new measure, which was named the *T-forward-risk-adjusted measure* by Jamshidian, prices of interest rate derivatives can be calculated as the discounted expected value of the payoff, which makes the calculation much simpler. However, explicitly determining this new measure can be complicated.

In this chapter we provide an alternative method to determine the T-forward-risk-adjusted measure for interest rate models. We do so by showing that the fundamental solutions to the pricing partial differential equation can be interpreted as the discounted probability density functions associated with the T-forward-risk-adjusted measure. A method to obtain fundamental solutions from the partial differential equation using Fourier transforms is introduced.

We define the class of normal models. These are interest rate models where the spot interest rate is a deterministic function of an underlying normally distributed stochastic process that drives the economy. We show that the models with the richest analytical structure belong to the class of normal models. These models with a rich analytical structure have also normally distributed fundamental solutions. Using the methods introduced in this chapter we derive an important theoretical result. We prove that within the class of normal models only the set of models where the spot interest rate is either a linear or a quadratic function of the underlying process has normally distributed fundamental solutions.

The rest of this chapter is organised as follows. In Section 4.1 we show how prices of interest rate derivatives can be calculated in a simpler fashion by changing to the T-forward-risk-adjusted measure. In Section 4.2 we show

how the fundamental solutions to the partial differential equation can be interpreted as the probability density functions under the new measure. A method for obtaining fundamental solutions is explained in Section 4.3 and an example is given in Section 4.4 that illustrates the concepts developed so far. Finally, in Section 4.5, we prove the theoretical result that within the class of normal models only the set of models where the spot interest rate is either a linear or a quadratic function of the underlying process has normally distributed fundamental solutions.

4.1 Forward-Risk-Adjusted Measure

In Chapter 3 we derived the result that prices of interest rate derivatives can be calculated by solving the appropriate partial differential equation (the Vasicek methodology) or by taking the expectation with respect to the equivalent martingale measure (the HJM methodology). The price $V(t, r)$ of an interest rate derivative can be expressed as

$$V(t, r) = \mathbb{E}^* \left(e^{-\int_t^T r(s)\, ds} H(T, r(T)) \mid \mathcal{F}_t \right), \qquad (4.1)$$

where the expectation is taken with respect to the equivalent martingale measure $\mathbb{Q}^*$ and $H(T, r)$ denotes the payoff of the derivative at time T.

When we determined the equivalent martingale measure in the HJM economy, we used the value of the money-market account $B(t)$ as a numeraire. However, this is not the only choice we could have made. As was argued in Chapter 2, it is possible to use any financial instrument with a strictly positive price (and no intermediate payouts) as a numeraire. The Change of Numeraire Theorem is very powerful in the context of valuing interest rate derivatives. Instead of using the value of the money-market account $B(t)$ as a numeraire, the prices of discount bonds $D(t, T)$ can also be used as a numeraire. A very convenient choice is to use the discount bond with maturity T as a numeraire for derivatives which have a payoff $H(T, r(T))$ at time T. If we denote the probability measure associated to the numeraire $D(t, T)$ by $\mathbb{Q}^T$ we can apply the Change of Numeraire Theorem as follows. Under the measure $\mathbb{Q}^T$ the prices $V(t, r)/D(t, T)$ are martingales for $t < T$. Hence, applying the definition of a martingale we obtain

$$\mathbb{E}^T \left(\frac{V(T, r(T))}{D(T, T)} \mid \mathcal{F}_t \right) = \frac{V(t, r)}{D(t, T)}. \qquad (4.2)$$

However, at time T the price of the discount bond $D(T, T) \equiv 1$ and the price of the instrument V is given by its payoff $V(T, r(T)) = H(T, r(T))$. So, the equation reduces to

$$V(t, r) = D(t, T) \mathbb{E}^T \left(H(T, r(T)) \mid \mathcal{F}_t \right). \qquad (4.3)$$

If we compare this expression with (4.1), we see that we have managed to express the expectation of the discounted payoff as a discounted expectation of the payoff. We have eliminated the problem of the correlation between the discounting term and the payoff term.

The measure $\mathbb{Q}^T$ has another very interesting property, which actually gave it the name T-forward-risk-adjusted measure. Under the T-forward-risk-adjusted measure, the forward rate $f(t, T)$ is equal to the expected value of the spot interest rate at time T. The following argument shows why this is true. A discount bond $D(t, T)$ has a payoff of 1 at time T. Using (4.1), the price of the discount bond can be expressed as

$$D(t, T) = E^* \left(e^{-\int_t^T r(s)\, ds} \cdot 1 \,\Big|\, \mathcal{F}_t \right). \tag{4.4}$$

Taking derivatives with respect to T yields

$$-\frac{\partial}{\partial T} D(t, T) = \mathbb{E}^* \left(e^{-\int_t^T r(s)\, ds} r(T) \,\Big|\, \mathcal{F}_t \right)$$

$$= D(t, T) \mathbb{E}^* \left(\frac{e^{-\int_t^T r(s)\, ds}}{D(t, T)} r(T) \,\Big|\, \mathcal{F}_t \right) \tag{4.5}$$

$$= D(t, T) \mathbb{E}^T \left(r(T) \,\big|\, \mathcal{F}_t \right),$$

where we have used the Change of Numeraire Theorem in the last step with Radon-Nikodym derivative

$$\frac{d\mathbb{Q}^T}{d\mathbb{Q}^*} = \frac{D(T, T)/D(t, T)}{B(T)/B(t)} = \frac{e^{-\int_t^T r(s)\, ds}}{D(t, T)}. \tag{4.6}$$

Using the definition of the forward rates $f(t, T) = -\partial/\partial T \log D(t, T)$ we can simplify this expression to

$$f(t, T) = \mathbb{E}^T \left(r(T) \,\big|\, \mathcal{F}_t \right) = \mathbb{E}^T \left(f(T, T) \,\big|\, \mathcal{F}_t \right), \tag{4.7}$$

which is the desired result. In other words, the instantaneous forward rate f is a martingale under the measure $\mathbb{Q}^T$.

Although we have eliminated in expression (4.3) the problem of the correlation between the discounting term and the payoff term, we have introduced the problem of having to determine the distribution of r under the probability measure $\mathbb{Q}^T$. In some cases it is possible to determine the probability measure directly from the Radon-Nikodym derivative; however, in general this can be complicated. In the next section we explore a different way of determining the distribution of r under the measure $\mathbb{Q}^T$ using the concept of fundamental solutions of the partial differential equation.

4.2 Fundamental Solutions

If we use the Vasicek methodology of Chapter 3, then the prices of interest rate derivatives are described by a partial differential equation of the form

$$V_t + \mu(t,r)V_r + \tfrac{1}{2}\sigma(t,r)^2 V_{rr} - rV = 0. \tag{4.8}$$

A partial differential equation of this form has the property that if two functions $V_1(t,x)$ and $V_2(t,x)$ both satisfy the same partial differential equation (disregarding any boundary conditions), then also the linear combination $a_1 V_1(t,x) + a_2 V_2(t,x)$ satisfies this partial differential equation. It is easily seen that this is true, due to the linearity of the differentiation operator. Even if we have an infinite parameterfamily of functions $V(t,x;y)$ that satisfy the same partial differential equation for every y, then the "infinite linear combination" $\int a(y)V(t,x;y)\,dy$ also satisfies this partial differential equation. (The interchange of differentiation and integration is allowed for sufficiently well-behaved functions a and V.) Suppose now, that we have a (double indexed) parameterfamily of functions $V^\delta(t,x;T,y)$ that satisfy a partial differential equation with boundary condition

$$V^\delta(T,x;T,y) = \delta(x-y) \tag{4.9}$$

for all y and all T. So, every function $V^\delta(t,x;T,y)$ collapses at time T into a Dirac delta-function[1] centred at point y. Consider now a boundary condition described by the function $H(T,x)$ at time T, then the function

$$V(t,x;T) = \int H(T,y)V^\delta(t,x;T,y)\,dy \tag{4.10}$$

satisfies the partial differential equation and satisfies the boundary condition at $t = T$. The last claim is true since at $t = T$ we have

$$\begin{aligned} V(T,x;T) &= \int H(T,y)V^\delta(T,x;T,y)\,dy \\ &= \int H(T,y)\delta(x-y)\,dy \\ &= H(T,x), \end{aligned} \tag{4.11}$$

where the last equality follows from the definition of the delta-function.

Hence, for *any* given boundary condition $H(T,x)$, the functions V^δ can be used to construct a solution to the partial differential equation. This is the reason why the functions V^δ are called *fundamental solutions* to this partial differential equation.

If the partial differential equation describes the development of prices through time in an economy (as, e.g. (4.8) does), we can give an economic

---

[1] For applications of delta-functions and fundamental solutions, see Griffel (1993) or Williams (1980).

interpretation for (4.10). The Dirac delta-function can be interpreted as the continuous equivalent of the payoff of an Arrow-Debreu security. Because $\delta(x - y) \neq 0$ only for $y = x$ and $\int \delta(x - y)\, dy = 1$, we could say that the delta-function gives a payoff worth 1 in the state of the world $y = x$. Hence, $V^\delta(t, x; T, y)$ can be viewed as the price at time t in state x of an Arrow-Debreu security that has a payoff of 1 at time T in state y. For discrete economies, it is well known that the price of any security with known payoffs at time T can be viewed as a portfolio of Arrow-Debreu securities and can be priced as the payoff-weighted sum over all states of the prices of the Arrow-Debreu securities. It is clear that (4.10) is the continuous equivalent of this payoff-weighted sum.

A discount bond D is a security that gives a payoff of 1 in all states of the world at maturity T. Hence, the price $D(t, T, y)$ at time t of a discount bond is $D(t, T, r) = \int V^\delta(t, r; T, z)\, dz$. If the economy is arbitrage-free, prices of Arrow-Debreu securities cannot become negative and prices of discount bonds are finite. These observations lead to another interpretation for (4.10). The function

$$p^\delta(t, r; T, z) = \frac{V^\delta(t, r; T, z)}{D(t, T, r)} \tag{4.12}$$

is non-negative and (by construction) integrates out to 1 with respect to z. Any function that is non-negative and integrates out to 1 can be interpreted as a probability density function, and (4.10) can be written as

$$V(t, r; T) = D(t, T, r) \int H(T, z) p^\delta(t, r; T, z)\, dz$$
$$= D(t, T, r) \widetilde{\mathbb{E}}^T\big(H(T, z) \mid \mathcal{F}_t\big), \tag{4.13}$$

where $\widetilde{\mathbb{E}}^T(. \mid \mathcal{F}_t)$ denotes the expectation operator, conditional on the information available at time t, with respect to the density $p^\delta(t, r; T, .)$.

If we combine this result with (4.1) and (4.3), where z represents $r(T)$, we get that

$$V(t, r; T) = \mathbb{E}^* \left(e^{-\int_t^T r(s)\, ds} H(T, z) \mid \mathcal{F}_t\right) = D(t, T, r) \widetilde{\mathbb{E}}^T\big(H(T, z) \mid \mathcal{F}_t\big). \tag{4.14}$$

We have expressed the expectation $\mathbb{E}^*$ of the discounted payoff, as the discounted expectation $\widetilde{\mathbb{E}}^T$ of the payoff, which is exactly the same result as in Section 4.1. Since this is true for any payoff H, the probability density functions $p^\delta(t, r; T, z)$ must be equal to the transition density functions $p^T(t, r; T, z)$ of the process r under the T-forward-risk-adjusted measure $\mathbb{Q}^T$.

4.3 Obtaining Fundamental Solutions

If we want to find fundamental solutions for the partial differential equation (4.8), we seek functions $V^\delta(t, r; T, z)$ that satisfy the partial differential equation, and collapse into a delta-function at $t = T$ for all T and all z. In order to facilitate the calculations, we will not try to solve for the fundamental solutions directly, but we will solve the partial differential equation for the *Fourier transform*[2] of the fundamental solutions, and obtain the fundamental solutions by inverting the Fourier transform.

Consider the function $\tilde{V}(t, r; T, \psi)$ defined as

$$\tilde{V}(t, r; T, \psi) = \int_{-\infty}^{\infty} e^{i\psi z} V^\delta(t, r; T, z)\, dz. \tag{4.15}$$

This function is the Fourier transform in the variable z of V^δ, where i is the imaginary number, for which $i^2 = -1$. The function $\tilde{V}$ satisfies the same partial differential equation as V^δ, but the boundary condition is given by

$$\tilde{V}(T, r; T, \psi) = \int_{-\infty}^{\infty} e^{i\psi z} \delta(r - z)\, dz = e^{i\psi r}, \tag{4.16}$$

where the last equality follows from the definition of the delta-function. We see that the boundary condition for $\tilde{V}$ is a very simple function.

The fundamental solutions V^δ can be obtained from $\tilde{V}$ by inverting the Fourier transform. This inversion is often simple, if we use the following property. Using (4.12) we can write the fundamental solution as the product of the discount bond price times a probability density function $V^\delta(t, r; T, z) = D(t, T, y)p^\delta(t, r; T, z)$. Hence, we can write $\tilde{V}$ as

$$\tilde{V}(t, r; T, \psi) = \int_{-\infty}^{\infty} e^{i\psi z} D(t, T, y)p^\delta(t, r; T, z)\, dz = D(t, T, y)\tilde{p}(t, r; T, \psi), \tag{4.17}$$

where $\tilde{p}$ is the Fourier transform of the density p^δ. However, the Fourier transform of a probability density function is the same as the *characteristic function* (see, e.g. Lukacs (1970)) of that density. For $\psi = 0$, the boundary condition of $\tilde{V}$ reduces to the boundary condition for a discount bond, and we get

$$D(t, T, r) = \tilde{V}(t, r; T, 0)$$

$$\tilde{p}(t, r; T, \psi) = \frac{\tilde{V}(t, r; T, \psi)}{\tilde{V}(t, r; T, 0)} = \frac{\tilde{V}(t, r; T, \psi)}{D(t, T, r)}. \tag{4.18}$$

---

[2] For applications of the Fourier transform, and for conditions under which the Fourier transform exists, see Duffie (1994). Heston (1993) uses a similar methodology to derive closed form expressions for a stochastic volatility model.

So, we see that both the price of a discount bond D, and the characteristic function $\tilde{p}$ of the density p^δ, can be obtained from $\tilde{V}$. Characteristic functions can be looked up in standard tables, or can be inverted numerically. Even without inverting the characteristic function, useful information about the distribution, like the moments or approximating distributions, can be obtained from the characteristic function. Once the discount bond price D and the densities p^δ have been obtained, prices for interest rate derivatives can be calculated by using (4.13).

4.4 Example: Ho-Lee Model

In this section we provide an example to illustrate the concepts developed so far. We consider the continuous-time limit of the Ho-Lee model we have already encountered in Chapter 3. For this model we can explicitly determine the T-forward-risk-adjusted measure both from the Radon-Nikodym derivative and from the fundamental solutions of the partial differential equation. We show that both derivations are consistent.

Other applications of the concepts developed in this chapter can be found in Chapter 5, where we analyse the Hull-White (1994) model, and in Chapter 6, where we analyse the one-factor squared Gaussian model.

4.4.1 Radon-Nikodym Derivative

In Chapter 3 we showed that, if we use the value of the money-market account $B(t)$ as a numeraire, under the equivalent martingale measure $\mathbb{Q}^*$ the spot interest rate r follows the process

$$dr = \theta^*(t)\, dt + \sigma\, dW^*, \tag{4.19}$$

where $\theta^*(t)$ is given by

$$\theta^*(t) = -\frac{\partial^2}{\partial t^2} \log D(0,t) + \sigma^2 t. \tag{4.20}$$

The prices at time $t = 0$ of interest rate derivatives with payoff $H(r(T))$ at maturity T can be calculated as

$$V_0 = E^* \left(e^{-\int_0^T r(s)\, ds} H(r(T)) \right). \tag{4.21}$$

Instead of using the value of the money-market account $B(t)$, we can also use the value of the T-maturity discount bond $D(t,T)$ as a numeraire. In Section 4.1 we explained that prices can also be calculated as

$$V_0 = D(0,T) E^T \left(H(r(T)) \right). \tag{4.22}$$

The Change of Numeraire Theorem provides us with the appropriate Radon-Nikodym derivative to change from $\mathbb{Q}^*$ to $\mathbb{Q}^T$

$$\frac{d\mathbb{Q}^T}{d\mathbb{Q}^*} = \frac{D(T,T)/D(0,T)}{B(T)/B(0)} = \frac{e^{-\int_0^T r(s)\,ds}}{D(0,T)}. \tag{4.23}$$

If we substitute the solution of the stochastic differential equation (4.19) into this equation and interchange the order of integration we obtain for $d\mathbb{Q}^T/d\mathbb{Q}^*$

$$\exp\left\{ -\log D(0,T) - \int_0^T \left(-\frac{\partial^2 \log D(0,s)}{\partial s^2} + \sigma^2 s \right)(T-s)\,ds \right.$$
$$\left. - \int_0^T \sigma(T-s)\,dW^*(s) \right\}. \tag{4.24}$$

If we work out the integrals, we see that the Radon-Nikodym derivative can be simplified to

$$\frac{d\mathbb{Q}^T}{d\mathbb{Q}^*} = \exp\left\{ -\tfrac{1}{6}\sigma^2 T^3 - \int_0^T \sigma(T-s)\,dW^*(s) \right\}. \tag{4.25}$$

From Girsanov's Theorem follows that a Radon-Nikodym derivative of this form can be obtained by setting $\kappa(t) = -\sigma(T-t)$. Hence, under the measure $\mathbb{Q}^T$ the process

$$W^T(t) = W^*(t) - \int_0^t (-\sigma)(T-s)\,ds \tag{4.26}$$

is also a Brownian motion. From this we obtain that for $t < T$ the spot interest rate r follows the process

$$dr = \left(\theta^*(t) - \sigma^2(T-t) \right) dt + \sigma\,dW^T \tag{4.27}$$

under the probability measure $\mathbb{Q}^T$.

4.4.2 Fundamental Solutions

In Chapter 3 we derived that prices of interest derivatives in the Ho-Lee model follow the partial differential equation

$$V_t + \theta^*(t)V_r + \tfrac{1}{2}\sigma^2 V_{rr} - rV = 0. \tag{4.28}$$

To find the function $\tilde{V}(t,r;T,\psi)$ for the Ho-Lee model, we have to solve (4.28) with respect to the boundary condition $\tilde{V}(T,r;T,\psi) = e^{i\psi r}$. As a solution we try

$$\tilde{V}(t,r;T,\psi) = \exp\{A(t;T\psi) + B(t;T,\psi)r\}. \tag{4.29}$$

Substituting this functional form in (4.28) and collecting terms yields

$$r(B_t - 1) + A_t + \theta^*(t)B + \tfrac{1}{2}\sigma^2 B^2 = 0, \tag{4.30}$$

which is solved if A and B solve the following system of ordinary differential equations

$$\begin{cases} B_t - 1 = 0 \\ A_t + \theta^*(t)B + \frac{1}{2}\sigma^2 B^2 = 0 \end{cases} \tag{4.31}$$

subject to the boundary conditions $A(T;T,\psi) = 0$ and $B(T;T,\psi) = i\psi$. The solutions for A and B are given by

$$B(t;T,\psi) = i\psi - (T - t)$$

$$A(t;T,\psi) = \frac{1}{6}\sigma^2(T - t)^3 - \int_t^T \theta^*(s)(T - s)\, ds \tag{4.32}$$

$$+ i\psi\left(\int_t^T \theta^*(s) - \sigma^2(T - s)\, ds\right) - \frac{1}{2}\psi^2\sigma^2(T - t).$$

If we substitute these expressions for A and B into (4.29) we obtain an explicit expression for the function $\widetilde{V}(t, r; T, \psi)$.

In Section 3 we showed that the price of a discount bond can be obtained by setting $\psi = 0$ in $\widetilde{V}$, this leads to

$$D(t,T,r) = \exp\left\{-r(T - t) + \frac{1}{6}\sigma^2(T - t)^3 - \int_t^T \theta^*(s)(T - s)\, ds\right\}. \tag{4.33}$$

This expression is consistent with the expression for the discount bond prices $D(0,T)$ derived in Chapter 3.

We have also shown in Section 3 that the characteristic function of the density p^δ can be obtained from $\widetilde{V}(t, r; T, \psi)/D(t, T, r)$, which yields

$$\exp\{i\psi M(t, T, r) - \frac{1}{2}\psi^2 \Sigma(t, T)\}, \tag{4.34}$$

where

$$M(t, T, r) = r + \int_t^T \left(\theta^*(s) - \sigma^2(T - s)\right) ds \tag{4.35}$$

$$\Sigma(t, T) = \sigma^2(T - t).$$

This characteristic function is the characteristic function of a normal distribution with mean $M(t, T, r)$ and variance $\Sigma(t, T)$. Hence, the fundamental solutions of the partial differential equation (4.28) imply that under the T-forward-risk-adjusted measure, the spot interest rate r has a normal distribution with mean $M(t, T, r)$ and variance $\Sigma(t, T)$. This result is consistent with the process (4.27) derived before.

4.5 Fundamental Solutions for Normal Models

In Chapter 3 we introduced the class of one-factor yield-curve models, where the spot interest rate r is described by the stochastic differential equation

$$dr = \mu(t, r)\, dt + \sigma(t, r)\, dW. \tag{4.36}$$

An important sub-class of models arises when the spot interest rate is modelled as follows

$$\begin{cases} dy = \big(\theta(t) - a(t)y\big)\, dt + \sigma(t)\, dW \\ r = F(t, y). \end{cases} \tag{4.37}$$

The first equation describes the dynamics of an underlying process y. The spot interest rate r is determined from the underlying process via the function $F(t, y)$. The stochastic differential equation defining y is a *linear equation in the narrow sense*.[3] Hence, the process y has a normal distribution. Therefore, we define this class of models as *normal models*. By making the appropriate choice for F, we can show that several well-known models fall in our class of normal models.

For $F(t, y) = y$, we obtain the Hull and White (1990a) model; the choice $F(t, y) = y^2$, leads to the one-factor version of the squared Gaussian model which is discussed in Chapter 6; and for $F(t, y) = e^y$, the Black and Karasinski (1991) model can be obtained. Finally, for the choices $a(t) \equiv 0$, $\sigma(t) \equiv \sigma$ and $F(t, r) = y$, we see that (4.37) reduces to the Ho-Lee model.

Using the methodology of Vasicek (1977), which is explained in Chapter 3, one can show that, for a normal model, the price $V(t, y; T)$ of an interest rate derivative security at time t which has a payoff at time T satisfies the partial differential equation

$$V_t + \mu^*(t, y)V_y + \tfrac{1}{2}\sigma(t)^2 V_{yy} - F(t, y)V = 0, \tag{4.38}$$

where

$$\mu^*(t, y) = \big(\theta(t) - a(t)y\big) - \lambda(t, y)\sigma(t) \tag{4.39}$$

and $\lambda(t, y)$ denotes the market price of risk. If one makes the additional assumption that the market price of risk $\lambda(t, y)$ is of the form (see also Hull and White (1990a), Footnote 1) $\lambda_1(t) + \lambda_2(t)y$, we can rewrite μ^* as

$$\mu^*(t, y) = \theta^*(t) - a^*(t)y, \tag{4.40}$$

where

$$\begin{aligned} \theta^*(t) &= \theta(t) - \lambda_1(t)\sigma(t) \\ a^*(t) &= a(t) + \lambda_2(t)\sigma(t). \end{aligned} \tag{4.41}$$

In this section we will investigate which models in the class of normal models have normally distributed fundamental solutions, since these models

---

[3] For a discussion of linear stochastic differential equations in the narrow sense, see Arnold (1992), Chapter 8.

are most likely to have a rich analytical structure. We will prove that the only normal models that have normally distributed fundamental solutions, are the models for which $F(t, y)$ is either a linear or a quadratic function in y. This result implies that only models like the Hull-White models, for which F is linear, and models for which F is quadratic have normally distributed fundamental solutions. Both models have a rich analytical structure. All other normal models, for example the Black and Karasinski (1991) model, for which F is an exponential function in y, have fundamental solutions that are *not* normally distributed.

In the remainder of this section we will prove the following theorem:

Theorem. *The only models in the class of normal models that have normally distributed fundamental solutions are models where $F(t, y)$ is either a linear or a quadratic function in y. These models will have a variance term independent of y, and a mean which is linear in y.*

Proof. The prices of interest rate derivatives in a normal model must satisfy the partial differential equation

$$V_t + \left(\theta^*(t) - a^*(t)y\right)V_y + \tfrac{1}{2}\sigma(t)^2 V_{yy} - F(t, y)V = 0. \qquad (4.42)$$

If a model has normally distributed fundamental solutions V^δ, then the Fourier transform $\widetilde{V}$ will be of the form

$$\widetilde{V}(t, y; T, \psi) = D(t, T, y)\exp\{i\psi M(t, T, y) - \tfrac{1}{2}\psi^2 \Sigma(t, T, y)\}, \qquad (4.43)$$

where $D(t, T, y)$ is the price of a discount bond, and M and Σ are the mean and variance of the normal distribution respectively. The price of a discount bond D must solve (4.42) with boundary condition $D(T, T, y) \equiv 1$.

Variance independent of y The function $\widetilde{V}$ defined in (4.43) must solve the partial differential equation (4.42). The partial derivatives of $\widetilde{V}$ are given by

$$\widetilde{V}_t = \widetilde{V}\left(\frac{D_t}{D} + i\psi M_t - \tfrac{1}{2}\psi^2 \Sigma_t\right)$$

$$\widetilde{V}_y = \widetilde{V}\left(\frac{D_y}{D} + i\psi M_y - \tfrac{1}{2}\psi^2 \Sigma_y\right)$$

$$\widetilde{V}_{yy} = \widetilde{V}\left(\frac{D_{yy}}{D} + i\psi\left(2\frac{D_y}{D}M_y + M_{yy}\right)\right.$$

$$\left. -\psi^2\left(M_y^2 + \frac{D_y}{D}\Sigma_y + \tfrac{1}{2}\Sigma_{yy}\right) - i\psi^3 M_y \Sigma_y + \tfrac{1}{4}\psi^4 \Sigma_y^2\right). \qquad (4.44)$$

If we substitute these partial derivatives into (4.42), we will get only one term containing ψ^4, namely $\tfrac{1}{2}\sigma(t)^2\widetilde{V}\tfrac{1}{4}\psi^4\Sigma_y^2$. However, the partial differential equation will only be solved for all ψ, if all terms containing ψ are identically equal to zero. Hence, the term containing ψ^4 can only be zero if $\Sigma_y \equiv 0$, but this implies that the variance term Σ must be independent of y.

Mean linear in y For Σ independent of y, the partial derivatives of $\tilde{V}$ reduce to

$$\tilde{V}_t = \tilde{V}\left(\frac{D_t}{D} + i\psi M_t - \tfrac{1}{2}\psi^2 \Sigma_t\right)$$

$$\tilde{V}_y = \tilde{V}\left(\frac{D_y}{D} + i\psi M_y\right) \qquad (4.45)$$

$$\tilde{V}_{yy} = \tilde{V}\left(\frac{D_{yy}}{D} + i\psi\left(2\frac{D_y}{D}M_y + M_{yy}\right) - \psi^2 M_y^2\right).$$

Substituting these partial derivatives into (4.42) and simplifying yields

$$-\tfrac{1}{2}\psi^2\left[\Sigma_t + \sigma(t)^2 M_y^2\right] + i\psi\left[M_t + (\theta^*(t) - a^*(t)y)M_y\right.$$
$$\left. + \tfrac{1}{2}\sigma(t)^2(2\tfrac{D_y}{D}M_y + M_{yy})\right] = 0. \quad (4.46)$$

Note that the other partial derivatives of D cancel against F, because the discount bond price is assumed to satisfy the partial differential equation. The term containing ψ^2 will be identically equal to zero only if $\Sigma_t + \sigma(t)^2 M_y^2 \equiv 0$, which is the case if M_y is a function of time only. Therefore M must be of the form

$$M(t,T,y) = M_0(t,T) + M_1(t,T)y, \qquad (4.47)$$

so we see that the mean M is linear in y.

F linear or quadratic in y For this linear form of M, we get that the coefficient of $i\psi$ in (4.46) will be identically zero only if

$$M_{0t} + M_{1t}y + (\theta^*(t) - a^*(t)y)M_1 + \sigma(t)^2\tfrac{D_y}{D}M_1 \equiv 0. \qquad (4.48)$$

This is possible only if D_y/D is either independent of y or a linear function in y. Hence, D is either of the form

$$D(t,T,y) = \exp\{A(t,T) + B(t,T)y\} \qquad (4.49)$$

or

$$D(t,T,y) = \exp\{A(t,T) + B(t,T)y + C(t,T)y^2\}. \qquad (4.50)$$

However, a discount bond price D must solve the partial differential equation (4.42). If $\log D$ is linear in y, then the partial derivative D_t/D is linear in y, and the derivatives D_y/D and D_{yy}/D are independent of y. In this case F can only be linear in y. On the other hand, if $\log D$ is quadratic in y, then the partial derivatives D_t/D and D_{yy}/D are quadratic in y, and the derivative D_y/D is linear in y. In this case F can only be quadratic in y. Hence, the only partial differential equations for which a normally distributed fundamental solution is feasible are the differential equations where F is either a linear or a quadratic function in y, which completes the proof.   □

5. The Hull-White Model

Given the tools we have developed in the previous chapters, we want to analyse some interest rate models which have a rich analytical structure.[4] In Chapter 4 we proved that only normal models where the spot interest rate is a linear or quadratic function of the underlying process y have normally distributed fundamental solutions. Hence, only these models are expected to have a rich analytical structure.

In this chapter we will analyse a model which is linear in the underlying process. It is the model proposed by Hull and White (1994). It is a generalisation of the continuous-time Ho-Lee model we have encountered before. The Ho-Lee model has a very rich analytical structure and is very easy to analyse; however, it does not provide a very realistic description of the behaviour of interest rates.

The first point on which the Ho-Lee model fails is the fact that it possesses no mean-reversion of interest rates. On the basis of economic theory, there are compelling arguments for the mean-reversion of interest rates. When rates are high, the economy tends to slow down and investments will decline. This implies there is less demand for money and rates will tend to decline. When rates are low, it is relatively cheap to invest, and rates will tend to rise. To obtain a more realistic model, Hull and White added mean-reversion to the Ho-Lee model.

The second weakness of the Ho-Lee model is the fact that the interest rates are normally distributed, which implies that the interest rates can become negative with positive probability. This problem is not solved in the Hull-White model because interest rates are also normally distributed in the Hull-White model. However, the probability that interest rates become negative is much smaller in the Hull-White than in the Ho-Lee model.

Despite this weakness, the Hull-White model has gained considerable popularity due to its analytical tractability. The Hull-White model can be fitted to the initial term-structure of interest rates analytically. Also prices for discount bonds and options on discount bonds can be valued analytically. The prices of frequently traded instruments like caps, floors, swaptions and options on coupon bearing bonds can be expressed in terms of options on discount bonds, and can therefore be valued analytically in the Hull-White model.

---

[4] The author would like to thank John Hull for comments and helpful suggestions.

In this chapter we provide a different derivation of the analytical fomulæ in the Hull-White model. Using the techniques developed in the previous chapters, we find fundamental solutions to the partial differential equation. From the fundamental solutions, we obtain formulæ for the prices of discount bonds. We furthermore obtain an explicit expression for the transition densities under the T-forward-risk-adjusted measure. Using these results we obtain closed form expressions for the prices of options on bonds and various other interest rate derivatives, which are frequently traded in the market.

To calculate prices for other, more complex, interest rate derivatives, the pricing partial differential equation has to be solved numerically. We develop an explicit finite difference method to solve the partial differential equation. This algorithm is different from the algorithm developed by Hull and White (1994).

The rest of the chapter is organised as follows. In Section 5.1 we introduce the model and find a transformation that simplifies the partial differential equation. In Section 5.2 we derive analytical formulæ for some interest rate derivatives. In Section 5.3 we show how the Hull-White model can be fitted to the initial term-structure of interest rates, and how an explicit finite difference algorithm can be set up efficiently to calculate prices for exotic options. We conclude in Section 5.4 with some examples, where we illustrate the convergence of the explicit finite difference algorithm and compare it to the algorithm of Hull and White (1994).

5.1 Spot Rate Process

Hull and White (1994) assume that the spot interest rate r follows the process

$$dr = \left(\theta(t) - ar\right) dt + \sigma \, dW, \tag{5.1}$$

where $\theta(t)$ is an arbitrary function of time and a and σ are constants. This model looks very much like the continuous-time Ho-Lee model, except for the additional term $-ar$ in the drift. For $a > 0$ this term adds the desired mean reverting property to the Ho-Lee model.

The Hull-White model can be cast into the framework of Chapter 4 as follows. We can rewrite the spot rate process as

$$\begin{cases} du = \left(\theta(t) - au\right) dt + \sigma \, dW \\ r = u \end{cases} \tag{5.2}$$

and we see that the Hull-White model is a normal model where the spot interest rate r is a linear function of the underlying process u.

5.1.1 Partial Differential Equation

From Chapters 3 and 4 we know that the value $V(t, r)$ of an interest rate derivative follows the partial differential equation

$$V_t + \mu^*(t, r)V_r + \tfrac{1}{2}\sigma^2 V_{rr} - rV = 0, \tag{5.3}$$

with

$$\mu^*(t, r) = \theta(t) - ar - \lambda(t, r)\sigma. \tag{5.4}$$

If one makes the additional assumption that the market price of risk $\lambda(t, r)$ is a function $\lambda(t)$ of time only, we can rewrite the partial differential equation as

$$V_t + \left(\theta^*(t) - ar\right)V_r + \tfrac{1}{2}\sigma^2 V_{rr} - rV = 0, \tag{5.5}$$

where $\theta^*(t) = \theta(t) - \lambda(t)\sigma$.

5.1.2 Transformation of Variables

Consider the following transformation of variables

$$y = r - \alpha(t)$$

$$\alpha(t) = e^{-at}\left(r_0 + \int_0^t e^{au}\theta^*(u)\, du\right). \tag{5.6}$$

We have chosen $\alpha(t)$ in such a way that $y(0) = 0$.

The price of any interest rate derivative security in terms of the new variable y can be written as $g(t, y)$. We can establish the following relationships between V and g

$$\begin{aligned}
V(t, r) &\equiv g(t, y) = g\big(t, r - \alpha(t)\big) \\
V_t &= g_t - \left(-a\alpha(t) + \theta^*(t)\right)g_y \\
V_r &= g_y \\
V_{rr} &= g_{yy}.
\end{aligned} \tag{5.7}$$

Substituting these relations into (5.5) and using $r = y + \alpha(t)$, the partial differential equation reduces to

$$g_t - ay\, g_y + \tfrac{1}{2}\sigma^2 g_{yy} - \big(y + \alpha(t)\big)g = 0. \tag{5.8}$$

This partial differential equation can be interpreted as the partial differential equation corresponding to an economy where, under the equivalent martingale measure $\mathbb{Q}^*$, the spot interest rate is generated by

$$\begin{cases} dy = -ay\, dt + \sigma\, dW^* \\ r(t) = y(t) + \alpha(t). \end{cases} \tag{5.9}$$

The advantage of the transformation of variables now becomes apparent. The stochastic process for the underlying variable y is determined only by the volatility parameters a and σ and is independent of the function $\alpha(t)$. The property makes our transformed model much easier to analyse than the original Hull and White (1994) model.

The process that y is assumed to follow is a very simple stochastic process known as an *Ornstein-Uhlenbeck process* (see Arnold (1992), Chapter 8 for properties of this stochastic process). Given a value $y(t)$ at any point in time t, the probability distribution of $y(T)$ for some future time $T > t$ is a normal distribution with mean

$$e^{-a(T-t)} y(t) \tag{5.10}$$

and variance

$$\frac{\sigma^2}{2a} \left(1 - e^{-2a(T-t)} \right). \tag{5.11}$$

The effect of the mean-reversion for $a > 0$ is reflected both in the mean and the variance of the process. For large T the mean of y falls to 0, and the variance tends to $\sigma^2/2a$. This is a much more realistic behaviour than the Ho-Lee model where the variance term $\sigma^2(T - t)$ becomes very large for large T.

For the remainder of this chapter we will work with the transformed model (5.9). By inverting the transformation of variables (5.6) the interested reader can, of course, always translate our results in terms of the original variables r and θ^*.

5.2 Analytical Formulæ

Prices of interest rate derivatives can be calculated in two ways. Using the Feynman-Kac formula we can express solutions to the partial differential equation (5.8) in terms of the boundary condition $H(T, y(T))$ at time T as

$$g(t,y) = \mathbb{E}^* \left(e^{-\int_t^T r(s)\,ds} H(T, y(T)) \mid \mathcal{F}_t \right), \tag{5.12}$$

where $\mathbb{E}^*$ is the expectation with respect to the process (5.9). As we showed in Chapter 4, the price of a derivative with a payoff at time T can be more conveniently evaluated using the T-forward-risk-adjusted measure $\mathbb{Q}^T$. The price can then be expressed as

$$g(t,y) = D(t,T,y) \mathbb{E}^T \left(H(T, y(T)) \mid \mathcal{F}_t \right), \tag{5.13}$$

where $D(t,T,y)$ denotes the price of a discount bond with maturity T at time t. A convenient way to determine the discount bond price $D(t,T,y)$ and the distribution of y under the T-forward-risk-adjusted measure is to use the Fourier transform $\tilde{g}$ of the fundamental solutions g^δ.

5.2.1 Fundamental Solutions

To find the Fourier transform $\tilde{g}$ for the Hull-White model, we have to solve (5.8) with respect to the boundary condition $\tilde{g}(T, y; T, \psi) = e^{i\psi y}$. From the proof of Chapter 4 we know for a model where the spot interest rate is a linear function of the underlying process the function $\tilde{g}$ must be of the form

$$\tilde{g}(t, y; T, \psi) = \exp\{A(t; T, \psi) + B(t; T, \psi)y\}. \tag{5.14}$$

Substituting this functional form into (5.8) and rearranging terms yields

$$y(B_t - aB - 1) + A_t + \tfrac{1}{2}\sigma^2 B^2 - \alpha(t) = 0, \tag{5.15}$$

which is solved if A and B solve the following system of ordinary differential equations

$$\begin{cases} B_t - aB - 1 = 0 \\ A_t + \tfrac{1}{2}\sigma^2 B^2 - \alpha(t) = 0 \end{cases} \tag{5.16}$$

subject to the boundary conditions $A(T; T, \psi) = 0$ and $B(T; T, \psi) = i\psi$. The solution for A and B is given by

$$B(t; T, \psi) = i\psi e^{-a(T-t)} - \frac{1 - e^{-a(T-t)}}{a}$$

$$\begin{aligned} A(t; T, \psi) = {} & \frac{\sigma^2}{2a^3}\left(a(T-t) - 2(1 - e^{-a(T-t)}) + \tfrac{1}{2}(1 - e^{-2a(T-t)})\right) \\ & - i\psi\frac{\sigma^2}{2a^2}(1 - e^{-a(T-t)})^2 - \tfrac{1}{2}\psi^2\frac{\sigma^2}{2a}(1 - e^{-2a(T-t)}) \\ & - \int_t^T \alpha(s)\,ds. \end{aligned} \tag{5.17}$$

Substituting these expressions for A and B into (5.14) yields

$$\tilde{g}(t, y; T, \psi) = \exp\{A(t, T) - B(t, T)y + i\psi M(t, T, y) - \tfrac{1}{2}\psi^2 \Sigma(t, T)\}, \tag{5.18}$$

with

$$\begin{aligned} A(t, T) = {} & \frac{\sigma^2}{2a^3}\Big(a(T-t) - 2(1 - e^{-a(T-t)}) \\ & + \tfrac{1}{2}(1 - e^{-2a(T-t)})\Big) - \int_t^T \alpha(s)\,ds \end{aligned}$$

$$B(t, T) = \frac{1 - e^{-a(T-t)}}{a} \tag{5.19}$$

$$M(t, T, y) = ye^{-a(T-t)} - \frac{\sigma^2}{2a^2}\left(1 - e^{-a(T-t)}\right)^2$$

$$\Sigma(t, T) = \frac{\sigma^2}{2a}\left(1 - e^{-2a(T-t)}\right).$$

In Chapter 4 we demonstrated that the Fourier transform $\tilde{g}$ is the product of the discount bond price and the characteristic function of the probability density function under the T-forward-risk-adjusted measure. Hence, the discount bond price is given by

$$D(t, T, y) = \exp\{A(t, T) - B(t, T)y\}. \tag{5.20}$$

The remaining terms in $\tilde{g}$ can be recognised as the characteristic function of a normal distribution with mean M and variance Σ. Hence, the probability density function $p^T(t, y; T, z)$ for $y(T)$ under the T-forward-risk-adjusted measure is a normal probability density function with mean $M(t, T, y)$ and variance $\Sigma(t, T)$. If we compare this mean and variance to the mean and variance of the process y given in (5.10) and (5.11), we see that the variances are the same, but that the mean has changed.

For $a = 0$ the Hull-White model reduces to the continuous-time Ho-Lee model. It is left to the reader to verify that all the results derived above reduce to the formulæ for the Ho-Lee model derived in Chapters 3 and 4 when we take the limit $a \to 0$.

5.2.2 Option Prices

Let us consider the price of a European call option on a discount bond. Let $C(t, T, \mathcal{T}, K, y)$ denote the price at time t of a call option that gives at time T the right to buy a discount bond with maturity $\mathcal{T}$ for a price K, with $t < T < \mathcal{T}$. Suppose that at time T the value of $y(T)$ is equal to z, then the payoff $H(T, z)$ of this option is equal to $\max\{D(T, \mathcal{T}, z) - K, 0\}$. The payout of the option is non-zero if

$$z < \frac{A(T, \mathcal{T}) - \log K}{B(T, \mathcal{T})}. \tag{5.21}$$

Under the T-forward-risk-adjusted measure $\mathbb{Q}^T$ the price of the call option can be expressed as

$$\mathbf{C}(t, T, \mathcal{T}, K, y) = \mathbb{E}^T\left(\max\{D(T, \mathcal{T}, y(T)) - K, 0\} \mid \mathcal{F}_t\right). \tag{5.22}$$

Given the expressions we have derived for the price of the discount bond and the probability distribution p^T of $y(T)$ under the T-forward-risk-adjusted measure, this expectation can be written as

$$\int_{-\infty}^{\frac{A - \log K}{B}} \frac{e^{A(T, \mathcal{T}) - B(T, \mathcal{T})z} - K}{\sqrt{2\pi \Sigma(t, T)}} \exp\{-\tfrac{1}{2}\tfrac{(z - M(t, T, y))^2}{\Sigma(t, T)}\} \, dz. \tag{5.23}$$

Some calculation will confirm that this integral can be expressed in terms of cumulative normal distribution functions $N(.)$ as follows:

$$\mathbf{C} = D(t, \mathcal{T}, y)N(h) - D(t, T, y)KN(h - \Sigma_P), \tag{5.24}$$

where
$$\Sigma_P = B(T, \mathcal{T})\sqrt{\Sigma(t, \mathcal{T})}$$
$$h = \frac{\log\big(D(t, \mathcal{T}, y)/D(t, T, y)K\big)}{\Sigma_P} + \tfrac{1}{2}\Sigma_P. \tag{5.25}$$

The value of a put option **P** can be derived in a similar fashion, and can be expressed as

$$\mathbf{P} = D(t, T, y)KN(-h + \Sigma_P) - D(t, \mathcal{T}, y)N(-h). \tag{5.26}$$

5.2.3 Prices for Other Instruments

With the analytic formulæ for discount bonds and call- and put-options on discount bonds, prices can be calculated for the following interest rate derivatives that are currently traded:

- **Coupon bearing bonds** can be priced as a portfolio of discount bonds: one discount bond for every coupon paid plus one discount bond for the principal.
- **Swaps** can be split in a fixed leg and a floating leg. The floating leg is always worth par[5], the fixed leg can be viewed as a portfolio of discount bonds.
- A **European option on a coupon bearing bond** is an option on a portfolio of discount bonds. Jamshidian has shown that this option can be decomposed as a portfolio of options on discount bonds. This is a non-trivial procedure, for references see e.g. Jamshidian (1989), Hull and White (1990a) or Hull (2000, Chapter 21).
- **Caps** and **floors** can also be decomposed as portfolios of puts and calls on discount bonds. This is explained in the appendix at the end of this chapter.
- Since the floating leg in a swap is always worth par, a **swaption** can be viewed as an option on the fixed leg with a par strike. Since the fixed leg is a portfolio of discount bonds, we see that Jamshidian's decomposition into a portfolio of options on discount bonds can be used again.

For valuing American-style instruments, and various other more complex interest rate derivatives no analytic formulæ are available. Prices of these instruments can be calculated numerically using a finite difference approach. This will be explained in the next section.

---

[5] Assuming, of course, that the next floating rate has not been set yet; if it has been set, the next floating rate payment is known, and can be viewed as a discount bond.

5.3 Implementation of the Model

In this section we show how the Hull-White model we are considering can be fitted to the initial term-structure of interest rates by choosing $\alpha(t)$, and hence $\theta^*(t)$, such that the initial discount bond prices $D(0,T)$ are priced correctly. To calculate prices for derivatives for which no analytic formulæ are available, we develop an explicit finite difference method for the Hull-White model.

This explicit finite difference algorithm is different from the algorithm of Hull and White (1994). The most important difference is that our explicit finite difference algorithm uses an analytical formula for $\alpha(t)$ that is obtained from fitting the model analytically to the initial term-structure of interest rates. Hull and White fit their tree numerically to the term-structure before they start to calculate prices.

5.3.1 Fitting the Model to the Initial Term-Structure

The initial term-structure of interest rates is given by the prices of the discount bonds $D(0,T)$ at time $t = 0$. Given the formula for the discount bond price we have to solve

$$\log D(0,T) = A(0,T). \tag{5.27}$$

Substituting the definition for A given in (5.19) and taking derivatives with respect to T and simplifying yields

$$\alpha(T) = -\frac{\partial}{\partial T} \log D(0,T) + \frac{\sigma^2}{2a^2} \left(1 - e^{-aT}\right)^2. \tag{5.28}$$

The same result can be derived in a more elegant way using the T-forward-risk-adjusted measure. In Chapter 4 we derived the result that the T-forward rate is the expected value of the spot interest rate at time T under the T-forward-risk-adjusted measure. Hence, we have

$$f(0,T) = \mathbb{E}^T\left(r(T)\right) = \mathbb{E}^T\left(y(T) + \alpha(T)\right). \tag{5.29}$$

Seen from time $t = 0$ the process $y(T)$ has a normal distribution with mean $M(0,T,0)$ under the measure $\mathbb{Q}^T$ and we obtain immediately

$$\alpha(T) = f(0,T) + \frac{\sigma^2}{2a^2}\left(1 - e^{-aT}\right)^2, \tag{5.30}$$

where we have used the expression for M given in (5.19).

5.3.2 Transformation of Variables

In Section 5.1 we derived a transformation of variables that made the process y independent of the initial term-structure of interest rates. In this section we show that with an additional transformation of variables we can make the partial differential equation independent of the function $\alpha(t)$. If we define the function $h(t, y)$ as

$$h(t, y) = e^{\int_t^T \alpha(u)\, du} g(t, y), \tag{5.31}$$

we get the following partial differential equation for h

$$h_t - ayh_y + \tfrac{1}{2}\sigma^2 h_{yy} - yh = 0 \tag{5.32}$$

and we see that this partial differential equation has no longer terms dependent on t.

Solving (5.32) numerically, one can calculate values for $h(t, y)$. The value $V(t, r)$ of an interest rate derivative security is obtained from $h(t, y)$ via

$$V(t, r) = e^{-\int_t^T \alpha(u)\, du} h\big(t, r - \alpha(t)\big). \tag{5.33}$$

Using the analytic formula for $\alpha(t)$ given in (5.28), the integral of α can be calculated as

$$\int_t^T \alpha(u)\, du = -\log \frac{D(0, T)}{D(0, t)}$$
$$+ \frac{\sigma^2}{2a^3}\left(a(T - t) - 2(e^{-at} - e^{-aT}) + \tfrac{1}{2}(e^{-2at} - e^{-2aT})\right). \tag{5.34}$$

5.3.3 Trinomial Tree

Suppose we construct a grid with steps Δy along the y-axis, and steps Δt along the t-axis. A node (i, j) on the grid is a point where $t = i\Delta t$ and $y = j\Delta y$. On this grid we can calculate in every node a value h_{ij}, that will be an approximation of the "true" value $h(i\Delta t, j\Delta y)$. In the explicit finite difference approach the partial derivatives of h are approximated as follows[6]

$$h_{yy} \approx \frac{1}{(\Delta y)^2}\{h_{i+1,j+1} - 2h_{i+1,j} + h_{i+1,j-1}\}$$
$$h_y \approx \frac{1}{2\Delta y}\{h_{i+1,j+1} - h_{i+1,j-1}\} \tag{5.35}$$
$$h_t \approx \frac{1}{\Delta t}\{h_{i+1,j} - h_{ij}\}$$

---

[6] For an introduction to the numerical solution of partial differential equations, see Hull and White (1990b) or Smith (1985).

To solve the partial differential equation numerically one imposes the restriction that in every node (i,j) the approximations of the partial derivatives satisfy the partial differential equation (5.32) exactly, so we get after solving for h_{ij}

$$h_{ij} = \frac{\Delta t}{1 + j\Delta y \Delta t} \left\{ \left(\frac{\sigma^2}{2(\Delta y)^2} - \tfrac{1}{2}aj \right) h_{i+1,j+1} \right.$$
$$+ \left(-\frac{\sigma^2}{(\Delta y)^2} + \frac{1}{\Delta t} \right) h_{i+1,j} \qquad (5.36)$$
$$\left. + \left(\frac{\sigma^2}{2(\Delta y)^2} + \tfrac{1}{2}aj \right) h_{i+1,j-1} \right\}.$$

If we set $\Delta y = \sigma\sqrt{3\Delta t}$, the equation reduces to

$$h_{ij} = \frac{1}{1 + j\Delta y \Delta t} \left\{ p_u h_{i+1,j+1} + p_m h_{i+1,j} + p_d h_{i+1,j-1} \right\} \qquad (5.37)$$

with

$$\begin{aligned} p_u &= \tfrac{1}{6} - \tfrac{1}{2}aj\Delta t \\ p_m &= \tfrac{2}{3} \\ p_d &= \tfrac{1}{6} + \tfrac{1}{2}aj\Delta t. \end{aligned} \qquad (5.38)$$

For $-\frac{1}{3}\frac{1}{a\Delta t} < j < \frac{1}{3}\frac{1}{a\Delta t}$, the numbers p_u, p_m and p_d are all positive and sum to 1, and can be interpreted as trinomial probabilities. For increasing values of $|j|$, the probability of jumping towards $y = 0$ increases. This reflects the mean reverting behaviour of the process $y(t)$. In order to prevent the trinomial probabilities from going negative, we cannot use a finite difference grid that is arbitrarily large. At some level $j^+ < \frac{2}{3}\frac{1}{a\Delta t}$, we want to express h_{ij^+} in terms of h_{i+1,j^+}, h_{i+1,j^+-1} and h_{i+1,j^+-2}. By doing so we avoid using h_{i+1,j^++1} and the grid will remain bounded at j^+. If we use the following approximations of the partial derivatives

$$h_{yy} \approx \frac{1}{(\Delta y)^2} \left\{ h_{i+1,j^+} - 2h_{i+1,j^+-1} + h_{i+1,j^+-2} \right\}$$
$$h_y \approx \frac{1}{2\Delta y} \left\{ 3h_{i+1,j^+} - 4h_{i+1,j^+-1} + h_{i+1,j^+-2} \right\} \qquad (5.39)$$
$$h_t \approx \frac{1}{\Delta t} \left\{ h_{i+1,j^+} - h_{ij^+} \right\},$$

we can express h_{ij^+} as

$$h_{ij^+} = \frac{1}{1 + j^+\Delta y \Delta t} \left\{ p_m^+ h_{i+1,j^+} + p_d^+ h_{i+1,j^+-1} + p_{dd}^+ h_{i+1,j^+-2} \right\} \qquad (5.40)$$

with

$$\begin{aligned} p_m^+ &= \tfrac{7}{6} - \tfrac{3}{2}aj^+\Delta t \\ p_d^+ &= -\tfrac{1}{3} + 2aj^+\Delta t \\ p_{dd}^+ &= \tfrac{1}{6} - \tfrac{1}{2}aj^+\Delta t. \end{aligned} \qquad (5.41)$$

These probabilities are all positive for $\frac{1}{6}\frac{1}{a\Delta t} < j^+ < \frac{1}{3}\frac{1}{a\Delta t}$. We can also bound the grid from below at a level $j^- > -\frac{1}{3}\frac{1}{a\Delta t}$. At j^- we get

$$h_{ij-} = \frac{1}{1 + j^-\Delta y\Delta t}\{p_{uu}^- h_{i+1,j^-+2} + p_u^- h_{i+1,j^-+1} + p_m^- h_{i+1,j^-}\} \qquad (5.42)$$

with probabilities

$$
\begin{aligned}
p_{uu}^- &= \tfrac{1}{6} + \tfrac{1}{2}aj^-\Delta t \\
p_u^- &= -\tfrac{1}{3} - 2aj^-\Delta t \\
p_m^- &= \tfrac{7}{6} + \tfrac{3}{2}aj^-\Delta t,
\end{aligned}
\qquad (5.43)
$$

which are all positive for $-\frac{1}{3}\frac{1}{a\Delta t} < j^- < -\frac{1}{6}\frac{1}{a\Delta t}$.

Instead of first calculating h_{ij} in every node, and then, calculating V_{ij} from h_{ij}, using (5.33), we can calculate prices V_{ij} more efficiently by rewriting the differencing scheme as

$$g_{ij} = \frac{e^{-\int_{i\Delta t}^{(i+1)\Delta t}\alpha(u)\,du}}{1 + j\Delta y\Delta t}\{p_u g_{i+1,j+1} + p_m g_{i+1,j} + p_d g_{i+1,j-1}\}, \qquad (5.44)$$

where we have used (5.31). The price V_{ij} of an instrument can be obtained from g_{ij} using $V(t,r) = g(t, r - \alpha(t))$, where $\alpha(t)$ can be evaluated using (5.28). For every i, the expression $\int \alpha(u)\,du$ can be calculated analytically using (5.34).

The finite difference method outlined above can be implemented as follows. For an instrument with maturity T, and a given number of steps N, we can calculate the step-sizes $\Delta t = T/N$ and $\Delta y = \sigma\sqrt{3\Delta t}$. We set $j^{max} = \lceil\frac{1}{6}\frac{1}{a\Delta t}\rceil$, which is the first integer value of j for which we can bound the grid at $j^+ = -j^- = j^{max}$ without creating negative probabilities. Then for $i = 0, \ldots, j^{max}$ we can build a normal trinomial tree with probabilities p_u, p_m and p_d. For $i = j^{max} + 1, \ldots, N$ we build a trinomial tree that jumps "inward" at $j = j^{max}$ and $j = -j^{max}$, where the adjusted probabilities p^+ and p^- have to be used. The value for a derivative can be calculated by filling $h_{N,j}$ with the payoff $H(N\Delta t, j\Delta y)$ and then calculating backward using the backward recursion formula (5.44). Due to the fact that the tree jumps inward we only need a boundary condition at time T; "upper" and "lower" boundary conditions do not have to be supplied.

5.4 Performance of the Algorithm

The analysis of the Hull-White model in this chapter is different from the analysis in the papers of Hull and White (1990a, 1994). However, the analytical formulæ we obtain for the prices of discount bonds and options on discount bonds are consistent with the results of Hull and White.

Table 5.1. Calculation times

Calculation times in milliseconds
on 486DX/66MHz computer
5yr option on 9yr discount bond
$a = 0.10, \sigma = 0.01$

Steps	European		American	
	HW	FD	HW	FD
10	4.0	4.0	11.1	8.2
20	8.4	7.8	23.8	17.0
30	14.7	12.5	39.6	27.9
40	22.3	17.7	57.2	39.8
50	32.2	23.4	78.7	53.2

Table 5.2. Prices (in bp) of put options

Put options on 9yr discount bond
$a = 0.10, \sigma = 0.01, Z(T) = 0.08 - 0.05e^{-0.18T}$

Mat/Strike	Steps	European		American*	
		HW	FD	HW	FD
$T = 3$yr	10	199	197	200	198
	20	196	195	197	196
	30	194	193	195	194
$K = 0.63$	40	193	193	194	194
	50	193	193	194	194
	Anal.	193			
$T = 5$yr	10	141	138	149	145
	20	139	138	147	145
	30	138	137	146	145
$K = 0.72$	40	138	137	145	145
	50	137	137	145	144
	Anal.	136			
$T = 7$yr	10	103	101	120	116
	20	100	99	117	115
	30	99	98	115	114
$K = 0.85$	40	98	98	115	114
	50	98	97	114	113
	Anal.	97			

* American style put option gives on early exercise
at time t the right to sell a discount bond with
maturity $t + 9 - T$ years for a price of K.

The explicit finite difference (FD) algorithm we developed in the previous section is different from the algorithm of Hull and White (1994) (HW). Due to the fact that Hull and White fit their tree numerically to the term-structure before they start to calculate prices, the HW algorithm is slower than the FD algorithm. This is confirmed in Table 5.1, where we report the calculation times needed for the two algorithms. We see that the FD algorithm is up to 40% faster than the HW algorithm.

The difference in calculation times between the two algorithms can be largely reduced by enhancing the original Hull and White (1994) algorithm by also fitting it analytically to the initial term-structure of interest rates.

Another difference between the FD algorithm and Hull and White's is the fact that we build a tree for the spot interest rate r, while Hull and White build a tree for the (continuous) Δt-interest rate. This means that all analytical formulæ have to be re-expressed in terms of the Δt rate[7] before they can be used to calculate prices in the Hull-White model.

In Table 5.2 we compare the convergence of both algorithms for European- and American-style put options with different maturities on a 9-year discount bond. The initial term-structure of interest rates is given via the zero-curve $Z(T)$. The prices of the discount bonds $D(0,T)$ can be calculated from the zero-curve via $D(0,T) = \exp\{-Z(T)T\}$. We see that both the HW algorithm and the FD algorithm converge very fast. Convergence to within 1 basispoint (bp) is reached with a tree that takes 50 time steps. The FD algorithm converges slightly faster than the HW algorithm, but the difference is not very large.

5.5 Appendix

In this appendix we show how the value of a cap or a floor can be expressed as a portfolio of puts or calls on discount bonds respectively.

As is explained in Hull (2000, Chapter 20) a cap contract with strike K on the 3-months interest rate is a portfolio of options on the quarterly interest payments that have to be made. The individual options are known as *caplets*. At the beginning of a quarter, the (discrete) 3-months LIBOR rate L_3 is observed. Hence, at the end of the quarter the writer of the cap has to make a payment of

$$N\Delta t \max\{L_3 - K, 0\}, \tag{5.45}$$

where N is the principal and Δt is the daycount fraction. For example, for 3-months US-dollar interest rates the Act/360 daycount convention is used. At the beginning of the quarter, this payoff has a value of

---

[7] The spot interest rate can be expressed in terms of the Δt-rate using the identity $D(t, t + \Delta t, r) = \exp\{-r_{\Delta t}\Delta t\}$, where $r_{\Delta t}$ denotes the continuously compounded Δt-interest rate.

$$N \frac{\Delta t}{1 + \Delta t L_3} \max\{L_3 - K, 0\}. \tag{5.46}$$

A few lines of algebra shows that this is equal to

$$N(1 + \Delta t K) \max\left\{ \frac{1}{1 + \Delta t K} - \frac{1}{1 + \Delta t L_3}, 0 \right\}. \tag{5.47}$$

The expression $1/(1 + \Delta t L_3)$ is equal to the value of the 3-months discount bond $D(T, \mathcal{T}, r)$, where T and $\mathcal{T}$ denote the start- and end-date of the 3-months period under consideration. Hence, the value of the caplet is equal to $N(1 + \Delta t K)$ times the value of a put option on the 3-months discount bond with strike $1/(1 + \Delta t K)$. The value of a floor can be derived using similar arguments.

6. The Squared Gaussian Model

After the analysis in Chapter 5 of the Hull-White model, where the spot interest rate is a linear function of the underlying process, we turn our attention to a model where the spot interest rate is a quadratic function of the underlying process.[8]

In Chapter 5 we showed that the Hull-White model has a rich analytical structure. Analytical formulæ for the prices of discount bonds and options on discount bonds can be obtained in this model. With these formulæ, prices for frequently traded instruments like caps, floors and options on coupon bonds can be calculated analytically. Also, efficient numerical procedures exist for calculating the prices of derivatives for which no analytical formulæ exist. A major disadvantage of the Hull-White approach, however, lies in the fact that negative interest rates can occur, due to the fact that interest rates are normally distributed.

This problem can be circumvented by assuming that the spot interest rate is a quadratic function of the underlying process. This type of models is known as *squared Gaussian models*.

In this chapter, we show that for the one-factor squared Gaussian model an analytical structure as rich as in the Hull-White model can be obtained, with the additional advantage that the interest rates never become negative. Squared Gaussian models were first studied by Beaglehole and Tenney (1991) and Jamshidian (1996), these papers show that squared Gaussian models possess some analytical structure.

Using a different approach, we are able to carry the analysis of the one-factor squared Gaussian model considerably further. We provide analytical formulæ for the prices of discount bonds and options on discount bonds, and we show how the model can be fitted analytically to the initial term-structure. In the one-factor squared Gaussian model, the prices of options on discount bonds can be expressed in terms of cumulative normal distribution functions. Also we show that prices of other derivatives can be calculated efficiently in this model using a trinomial tree.

The rest of the chapter is organised as follows. In Section 6.1 we introduce the model and find a transformation that simplifies the partial differential equation. How we obtain fundamental solutions is explained in Section 6.2,

---

[8] This chapter is based on Pelsser (1997).

where we also derive analytical formulæ for some interest rate derivatives. In Section 6.3 we show how our model can be fitted to the initial term-structure, and how a trinomial tree can be set up efficiently to calculate prices for exotic options. We conclude with some examples, illustrating the convergence of the trinomial tree algorithm.

6.1 Spot Rate Process

In this chapter we analyse the one-factor squared Gaussian model

$$
\begin{cases}
du = \big(\theta(t) - au\big)\, dt + \sigma\, dW \\
r = u^2
\end{cases}
\tag{6.1}
$$

where a and σ are constants, and $\theta(t)$ is an arbitrary function of time. This is a normal model where the spot interest rate r is given by the square of the underlying process, which ensures that the spot interest rate never goes negative. For the general multi-factor model Jamshidian (1996) explains how prices of discount bonds can be obtained numerically and how the multi-factor model can be fitted to the initial term-structure by numerically solving a system of ordinary differential equations. Furthermore, he demonstrates that the prices of options on discount bonds can be expressed in terms of non-central chi-square distribution functions. Jamshidian's analysis for the multi-factor model is valid for this model also. However, using a different approach, we will provide analytical formulæ for the prices of discount bonds and we demonstrate that prices of interest rate derivatives can be expressed in terms of cumulative normal distribution functions.

6.1.1 Partial Differential Equation

As was shown in Chapters 3 and 4, the price $V(t, u)$ at time t of an interest rate derivative security satisfies the partial differential equation

$$
V_t + \big(\theta^*(t) - au\big)V_u + \tfrac{1}{2}\sigma^2 V_{uu} - u^2 V = 0,
\tag{6.2}
$$

where $\theta^*(t) = \theta(t) - \lambda(t)\sigma$. To make this partial differential equation easier to solve we employ a transformation of variables similar to the one used in Chapter 5

$$
y = u - \alpha(t)
$$
$$
\alpha(t) = e^{-at}\left(\sqrt{r_0} + \int_0^t e^{as}\theta^*(s)\, ds\right).
\tag{6.3}
$$

The price of any interest rate derivative security in terms of the new variable y can be written as $h(t, y) = V\big(t, y + \alpha(t)\big)$. In terms of y all prices follow the partial differential equation

$$h_t - ayh_y + \tfrac{1}{2}\sigma^2 h_{yy} - \big(y + \alpha(t)\big)^2 h = 0. \tag{6.4}$$

The transformed equation (6.4) can be interpreted as the partial differential equation belonging to an economy where, under the equivalent martingale measure $\mathbb{Q}^*$, the spot interest rate r is generated via

$$\begin{cases} dy = -ay\,dt + \sigma\,dW^* \\[2mm] r = \big(y + \alpha(t)\big)^2 \end{cases} \tag{6.5}$$

Given a value $y(t)$ at any point in time t, the probability distribution of $y(T)$ for some future time $T > t$ is a normal distribution with mean

$$e^{-a(T-t)}y(t) \tag{6.6}$$

and variance

$$\frac{\sigma^2}{2a}\left(1 - e^{-2a(T-t)}\right). \tag{6.7}$$

From a pricing point of view, the economy (6.1) with market price of risk $\lambda(t)$ and the "risk-neutral" economy (6.5) are indistinguishable, because the only observed variable in both economies is the spot interest rate r which has the same stochastic behaviour in both economies. It will make no difference for the price of a derivative whether we view it as a function of u or y. Therefore, we use in the remainder of this chapter the transformed partial differential equation (6.4), which corresponds to the risk-neutral economy (6.5).

6.2 Analytical Formulæ

Prices of interest rate derivatives can be calculated in two ways. Using the Feynman-Kac formula we can express solutions to the partial differential equation (6.4) in terms of the boundary condition $H(T, y(T))$ at time T as

$$h(t, y) = \mathbb{E}^* \left(e^{-\int_t^T r(s)\,ds} H(T, y(T)) \mid \mathcal{F}_t\right), \tag{6.8}$$

where $\mathbb{E}^*$ is the expectation with respect to the process (6.5). As we showed in Chapter 4, the price of a derivative with a payoff at time T can be more conveniently evaluated using the T-forward-risk-adjusted measure $\mathbb{Q}^T$. The price can then be expressed as

$$h(t, y) = D(t, T, y)\mathbb{E}^T \left(H(T, y(T)) \mid \mathcal{F}_t\right), \tag{6.9}$$

where $D(t, T, y)$ denotes the price of a discount bond with maturity T at time t. A convenient way to determine the discount bond price $D(t, T, y)$ and the distribution of y under the T-forward-risk-adjusted measure is to use the Fourier transform $\hat{h}$ of the fundamental solutions h^δ.

6.2.1 Fundamental Solutions

To find the function $\tilde{h}$ for the squared Gaussian model, we have to solve (6.4) with respect to the boundary condition $\tilde{h}(T, y; T, \psi) = e^{i\psi y}$. From the proof of Chapter 4 we know that, for a model where the spot interest rate is a quadratic function of the underlying process, the function $\tilde{h}$ must be of the form

$$\tilde{h}(t, y; T, \psi) = \exp\{A(t; T, \psi) - B(t; T, \psi)y - C(t; T, \psi)y^2\}. \qquad (6.10)$$

A function $\tilde{h}$ of this form satisfies (6.4) and the boundary condition $e^{i\psi y}$ if $A(t; T, \psi)$, $B(t; T, \psi)$ and $C(t; T, \psi)$ satisfy the following system of ordinary differential equations

$$\begin{cases} C_t - 2aC - 2\sigma^2 C^2 + 1 = 0 \\ B_t - aB - 2\sigma^2 BC + 2\alpha(t) = 0 \\ A_t + \frac{1}{2}\sigma^2 B^2 - \sigma^2 C - \alpha(t)^2 = 0 \end{cases} \qquad (6.11)$$

with $A(T; T, \psi) = C(T; T, \psi) = 0$ and $B(T; T, \psi) = -i\psi$. The solution to this system, with respect to the boundary conditions, is given by

$$\begin{aligned}
C(t; T, \psi) &= C(t, T) \\
B(t; T, \psi) &= B(t, T) - i\psi E(t, T) \\
A(t; T, \psi) &= A(t, T) + \int_t^T \frac{1}{2}\sigma^2\left(-2i\psi E(s, T)B(s, T) - \psi^2 E(s, T)^2\right) ds,
\end{aligned}$$
$$\qquad (6.12)$$

where

$$\begin{aligned}
E(t, T) &= \frac{2\gamma e^{\gamma(T-t)}}{(a + \gamma)e^{2\gamma(T-t)} + (\gamma - a)} \\
C(t, T) &= \frac{e^{2\gamma(T-t)} - 1}{(a + \gamma)e^{2\gamma(T-t)} + (\gamma - a)} \\
B(t, T) &= 2\int_t^T \frac{e^{\gamma s}\left((a + \gamma)e^{2\gamma(T-s)} + (\gamma - a)\right)}{e^{\gamma t}\left((a + \gamma)e^{2\gamma(T-t)} + (\gamma - a)\right)}\alpha(s)\, ds \\
A(t, T) &= \int_t^T \frac{1}{2}\sigma^2 B(s, T)^2 - \sigma^2 C(s, T) - \alpha(s)^2\, ds \\
\gamma &= \sqrt{a^2 + 2\sigma^2}.
\end{aligned} \qquad (6.13)$$

Substituting $A(t; T, \psi)$, $B(t; T, \psi)$ and $C(t; T, \psi)$ into (6.10) yields

$$\begin{aligned}
\tilde{h}(t, y; T, \psi) = \exp\big\{ A(t, T) - B(t, T)y - C(t, T)y^2 \\
+ i\psi M(t, T, y) - \tfrac{1}{2}\psi^2 \Sigma(t, T)\big\},
\end{aligned} \qquad (6.14)$$

with

$$M(t,T,y) = E(t,T)y - \int_t^T \sigma^2 E(s,T)B(s,T)\,ds$$

$$\Sigma(t,T) = \int_t^T \sigma^2 E(s,T)^2\,ds = \sigma^2 C(t,T).$$

(6.15)

For $\psi = 0$ the boundary condition for $\tilde{h}$ reduces to the boundary condition of the discount bond. Hence, the price of a discount bond is given by

$$D(t,T,y) = \tilde{h}(t,y;T,0) = \exp\{A(t,T) - B(t,T)y - C(t,T)y^2\}. \quad (6.16)$$

The terms containing ψ in (6.14) are easily recognised as the characteristic function of a normal distribution (see, e.g., Lukacs (1970)). The probability density functions p^T of y under the T-forward-risk-adjusted measure are, therefore, equal to a normal probability density function

$$p^T(t,y;T,z) = \frac{1}{\sqrt{2\pi\,\Sigma(t,T)}}\exp\left\{-\frac{1}{2}\frac{(z - M(t,T,y))^2}{\Sigma(t,T)}\right\}, \quad (6.17)$$

with mean $M(t,T,y)$ and variance $\Sigma(t,T)$. Note, that this mean and variance are different from the mean and variance of the process y given in (6.6) and (6.7).

Once we have explicitly determined prices for discount bonds, and have obtained the distribution of y under the T-forward-risk-adjusted measure, the price of an interest rate derivative with payoff $H(T,y(T))$ at time T can be calculated via (6.9) as

$$h(t,y;T) = D(t,T,y)\mathbb{E}^T\left(H(T,y(T))\mid \mathcal{F}_t\right)$$

$$= D(t,T,y)\int_{-\infty}^{+\infty} H(T,z)p^T(t,y;T,z)\,dz, \quad (6.18)$$

where z goes through all the possible values of $y(T)$.

6.2.2 Option Prices

Let us consider the price of a European call option on a discount bond. Let $C(t,T,\mathcal{T},K,y)$ denote the price at time t of a call option that gives at time T the right to buy a discount bond with maturity $\mathcal{T}$ for a price K, with $t < T < \mathcal{T}$. Suppose that at time T the value of $y(T)$ is equal to z, then the payoff $H(z)$ of this option is equal to $\max\{D(T,\mathcal{T},z) - K, 0\}$. The payout of the option is non-zero if

$$\log D(T,\mathcal{T},z) = A(T,\mathcal{T}) - B(T,\mathcal{T})z - C(T,\mathcal{T})z^2 > \log K. \quad (6.19)$$

This is true if (note that $C(T,\mathcal{T})$ is positive for all $T < \mathcal{T}$)

$$l = \frac{-B(T,\mathcal{T}) - \sqrt{d}}{2C(T,\mathcal{T})} < z < \frac{-B(T,\mathcal{T}) + \sqrt{d}}{2C(T,\mathcal{T})} = h, \quad (6.20)$$

where

$$d = B(T, \mathcal{T})^2 + 4C(T, \mathcal{T})\big(A(T, \mathcal{T}) - \log K\big), \tag{6.21}$$

provided that the discriminant d is positive. If $d \leq 0$, the payoff of the call option is never positive, and hence the call option price $\mathbf{C}$ is equal to zero. If $d > 0$, we can calculate the price $\mathbf{C}$ by integrating (6.18) over the region $l < z < h$. The integral can be expressed in terms of cumulative normal distribution functions $N(.)$ as follows

$$\mathbf{C}(t, T, \mathcal{T}, K, y) = D(t, \mathcal{T}, y) \left[N\left(\frac{h\tau - \nu}{\sqrt{\tau \Sigma(t, T)}}\right) - N\left(\frac{l\tau - \nu}{\sqrt{\tau \Sigma(t, T)}}\right) \right]$$

$$- D(t, T, y) K \left[N\left(\frac{h - M(t, T, y)}{\sqrt{\Sigma(t, T)}}\right) - N\left(\frac{l - M(t, T, y)}{\sqrt{\Sigma(t, T)}}\right) \right] \tag{6.22}$$

with

$$\nu = M(t, T, y) - B(T, \mathcal{T}) \Sigma(t, T)$$
$$\tau = 1 + 2C(T, \mathcal{T}) \Sigma(t, T). \tag{6.23}$$

The price for a put-option on a discount bond can be derived in a similar fashion.

For valuing American-style instruments, and various other exotic interest rate derivatives no analytic formulæ are available. Prices of these instruments have to be calculated numerically.

6.3 Implementation of the Model

As was stated before, yield-curve models take the initial term-structure of interest rates as an input, and price all interest rate derivatives off this curve using no-arbitrage arguments. In this section we show how the model we are considering can be fitted to the initial yield-curve by choosing $\alpha(t)$, and hence $\theta^*(t)$, such that the initial discount bond prices $D(0, T)$ are priced correctly. To calculate prices for derivatives for which no analytic formulæ are available, we provide a special trinomial tree approach for the squared Gaussian model.

6.3.1 Fitting the Model to the Initial Term-Structure

In Chapter 4 we proved the relation

$$f(t, T, y) = \mathbb{E}^T \big(r(T) \mid \mathcal{F}_t \big). \tag{6.24}$$

Hence, under the T-forward-risk-adjusted measure, the instantaneous forward rate for date T is equal to the expected value of the spot interest rate at time T. The spot interest rate is defined as $r = (y + \alpha)^2$, therefore we can express (6.24) as

$$f(t, T, y) = \Sigma(t, T) + \big(M(t, T, y) + \alpha(T)\big)^2, \qquad (6.25)$$

because, under the T-forward-risk-adjusted measure, y is normally distributed with mean M and variance Σ.

At time $t = 0$ the prices of the forward rates $f(0, T)$ are known for all T and $\alpha(T)$ has to be chosen such that these initial discount bond prices are priced correctly by the model. The value of $\alpha(T)$ can be related to $f(0, T)$ using (6.25), and we obtain

$$\begin{aligned} f(0, T) &= \Sigma(0, T) + \big(M(0, T, 0) + \alpha(T)\big)^2 \\ &= \Sigma(0, T) + \Big(-\int_0^T \sigma^2 E(s, T) B(s, T) \, ds + \alpha(T)\Big)^2, \end{aligned} \qquad (6.26)$$

where we have substituted the definition for M given in (6.15). If $f(0, T) \geq \Sigma(0, T)$, we can define

$$F(T) = \sqrt{f(0, T) - \Sigma(0, T)}, \qquad (6.27)$$

and get for $\alpha(T)$

$$\alpha(T) - \int_0^T \sigma^2 E(s, T) B(s, T) \, ds = F(T). \qquad (6.28)$$

The function $B(t, T)$ depends also on α, hence, (6.28) is an integral equation in α. In Appendix A of this chapter we show that the solution to (6.28) can be expressed as

$$\alpha(T) = F(T) + 2e^{-aT} \int_0^T e^{as} \Sigma(0, s) F(s) \, ds. \qquad (6.29)$$

It is clear that for $f(0, T, 0) < \Sigma(0, T)$, the model cannot be fitted to the initial term-structure.

Once $\alpha(t)$ is determined from the initial term-structure, prices for discount bonds and options on discount bonds can be calculated. All pricing formulæ depend on the functions $A(t, T)$ and $B(t, T)$, which can be determined from (6.13). However, calculating A and B from (6.13) would involve a numerical integration for every value of $A(t, T)$ or $B(t, T)$ needed. As was shown by Jamshidian (1996), $A(t, T)$ and $B(t, T)$ can be evaluated more efficiently, by expressing them in terms of $D(0, T, 0)$ and $B(0, T)$. The values for $D(0, T, 0)$ are known at $t = 0$ for all T, and the values for $B(0, T)$ have to be calculated only once for different values of T and can then be stored. We show in Appendix B of this chapter how this procedure can be implemented for our model.

6.3.2 Trinomial Tree

For interest rate derivatives for which no analytical formulæ exist, the partial differential equation (6.4) has to be solved numerically. In the remainder of this section we show how the partial differential equation can be solved using an explicit finite difference method. Then we show how the explicit finite difference method can be implemented efficiently as a trinomial tree. We conclude with an example to illustrate the convergence of the trinomial tree algorithm.

In Chapter 5 we derived an explicit finite difference algorithm for the Hull-White model. We can use the same methodology to derive an explicit finite difference algorithm for the squared Gaussian model. If we choose a grid with spacing Δt and $\Delta y = \sigma\sqrt{3\Delta t}$ we obtain the following backward recursion formula

$$h_{ij} = \frac{1}{1+(j\Delta y+\alpha(i\Delta t))^2\Delta t}\left\{p_u h_{i+1,j+1} + p_m h_{i+1,j} + p_d h_{i+1,j-1}\right\} \qquad (6.30)$$

with

$$\begin{aligned} p_u &= \tfrac{1}{6} - \tfrac{1}{2}aj\Delta t \\ p_m &= \tfrac{2}{3} \\ p_d &= \tfrac{1}{6} + \tfrac{1}{2}aj\Delta t. \end{aligned} \qquad (6.31)$$

The only difference between this formula, and the recursion scheme of Chapter 5 is the first term. This term reflects the fact that in the squared Gaussian model the spot interest rate is a quadratic function of the underlying process y. Keeping this difference in mind, one can implement a trinomial tree for the squared Gaussian model in exactly the same way as we did for the Hull-White model in Chapter 5.

To analyse the convergence of the trinomial tree algorithm outlined above, we show in Table 6.1 prices of European- and American-style put options on a 9 year discount bond, for different number of steps. The initial term-structure of interest rates is given via the zero-curve $Z(T)$. The prices of the discount bonds $D(0,T)$ can be calculated from the zero-curve via $D(0,T) = \exp\{-Z(T)T\}$. It is clear from this table, that the trinomial algorithm converges very fast. Convergence within 1 basispoint (bp) is reached within 100 steps, both for the European- and the American-style options.

6.4 Appendix A

In this appendix we show how the integral equation (6.28) can be solved. Using (6.27) for the right-hand side of (6.28), and using the definitions for E and B given in (6.13), we get

Table 6.1. Prices (in bp) of put options.

Put options on 9yr discount bond
$a = 0.10, \sigma = 0.03, Z(T) = 0.08 - 0.05e^{-0.18T}$

Mat/Strike	Steps	European	American*
$T = 3\text{yr}$	20	164	168
	40	161	165
	60	159	164
	80	160	165
$K = 0.6$	100	160	165
	Anal.	160	
$T = 5\text{yr}$	20	159	179
	40	156	177
	60	154	176
	80	154	175
$K = 0.7$	100	153	175
	Anal.	153	
$T = 7\text{yr}$	20	154	197
	40	151	194
	60	150	193
	80	149	193
$K = 0.85$	100	149	192
	Anal.	148	

* American style put option gives on early exercise
at time t the right to sell a discount bond
with maturity $t + 9 - T$ yr for a price K.

$$\alpha(T) - \sigma^2 \int_{s=0}^{T} \int_{u=s}^{T} 2 \frac{2\gamma e^{\gamma(T-s)}}{(a+\gamma)e^{2\gamma(T-s)} + (\gamma - a)}$$

$$\times \frac{e^{\gamma u}\left((a+\gamma)e^{2\gamma(T-u)} + (\gamma - a)\right)}{e^{\gamma s}\left((a+\gamma)e^{2\gamma(T-s)} + (\gamma - a)\right)} \alpha(u)\, du\, ds = F(T). \quad (6.32)$$

After interchanging the order of integration, we can simplify this expression
to

$$\alpha(T) - \frac{2\sigma^2 e^{\gamma T}}{(a+\gamma)e^{2\gamma T} + (\gamma - a)} \int_0^T (e^{\gamma u} - e^{-\gamma u})\alpha(u)\, du = F(T). \quad (6.33)$$

Readers familiar with literature on integral equations will recognise (6.33) as
a linear Volterra integral equation of the second kind

$$\alpha(T) - \int_0^T K(T, u)\alpha(u)\, du = F(T), \quad (6.34)$$

with a separable integration kernel $K(T, u)$ equal to

$$K(T, u) = \frac{2\sigma^2 e^{\gamma T}}{(a + \gamma)e^{2\gamma T} + (\gamma - a)}(e^{\gamma u} - e^{-\gamma u}).$$ (6.35)

It is well known (see Hochstadt (1973) or Griffel (1993)) that a linear Volterra integral equation with a continuous and bounded kernel has a unique solution for every continuous function $F(T)$. In this case, where the kernel is separable, the integral equation (6.33) can be solved as follows. After differentiating (6.33) with respect to T, we obtain

$$\alpha_T(T) - 2\sigma^2 \frac{\gamma e^{\gamma T}((a+\gamma)e^{2\gamma T} + (\gamma - a)) - e^{\gamma T}(a+\gamma)e^{2\gamma T}2\gamma}{((a+\gamma)e^{2\gamma T} + (\gamma - a))^2} \int_0^T (e^{\gamma u} - e^{-\gamma u})\alpha(u)\,du$$

$$- \frac{2\sigma^2 e^{\gamma T}}{(a + \gamma)e^{2\gamma T} + (\gamma - a)}(e^{\gamma T} - e^{-\gamma T})\alpha(T) = F_T(T).$$ (6.36)

Rewriting (6.33), the remaining integral can be expressed as

$$\int_0^T (e^{\gamma u} - e^{-\gamma u})\alpha(u)\,du = (\alpha(T) - F(T))\frac{(a + \gamma)e^{2\gamma T} + (\gamma - a)}{2\sigma^2 e^{\gamma T}}.$$ (6.37)

Substituting this expression into (6.36) yields, after some calculation

$$\alpha_T(T) + a\alpha(T) = \beta(T)F(T) + F_T(T),$$ (6.38)

where

$$\beta(T) = \gamma - \frac{2\gamma(\gamma - a)}{(a + \gamma)e^{2\gamma T} + (\gamma - a)}.$$ (6.39)

This ordinary differential equation in α has to be solved subject to the boundary condition $\alpha(0) = F(0)$. The solution can be expressed as

$$\alpha(T) = e^{-aT}\left(F(0) + \int_0^T e^{as}\beta(s)F(s)\,ds + \int_0^T e^{as}F_s(s)\,ds\right).$$ (6.40)

Using partial integration, the second integral can be rewritten as

$$\int_0^T e^{as}F_s(s)\,ds = e^{aT}F(T) - F(0) - \int_0^T ae^{as}F(s)\,ds,$$ (6.41)

and we obtain for (6.40)

$$\alpha(T) = F(T) + e^{-aT}\int_0^T e^{as}(\beta(s) - a)F(s)\,ds.$$ (6.42)

Substituting for β and simplifying, we end up with

$$\alpha(T) = F(T) + 2e^{-aT}\int_0^T e^{as}\sigma^2 \frac{e^{2\gamma s} - 1}{(a + \gamma)e^{2\gamma s} + (\gamma - a)}F(s)\,ds,$$ (6.43)

which is equal to expression (6.29).   □

6.5 Appendix B

In this appendix we show how values for $A(T, \mathcal{T})$ and $B(T, \mathcal{T})$ can be calculated efficiently, using methods outlined in Jamshidian (1996).

If an investor sells a discount bond with maturity $\mathcal{T}$ at an earlier time T, he will receive $D(T, \mathcal{T}, y(T))$. Therefore, for $t < T < \mathcal{T}$, one can view the discount bond $D(t, \mathcal{T}, y)$ as a security that has a payoff of $D(T, \mathcal{T}, y(T))$ at time T. Hence, we can write the price of a discount bond $D(t, \mathcal{T}, y)$ as

$$D(t, \mathcal{T}, y) = D(t, T, y)\mathbb{E}^{T}\left(D(T, \mathcal{T}, y(T)) \mid \mathcal{F}_t\right). \tag{6.44}$$

Under the T-forward-risk-adjusted measure, the random variable $y(T)$, seen from time t, is normally distributed with mean $M(t, T, y)$ and variance $\Sigma(t, T)$. Using the formula for the price of a discount bond given in (6.16), we can evaluate (6.44) as

$$\exp\{A(t, \mathcal{T}) - B(t, \mathcal{T})y - C(t, \mathcal{T})y^2\} = \exp\{A(t, T) - B(t, T)y - C(t, T)y^2\}$$
$$\times \frac{\exp\left\{A(T, \mathcal{T}) + \frac{\frac{1}{2}B(T,\mathcal{T})^2\Sigma(t,T) - B(T,\mathcal{T})M(t,T,y) - C(T,\mathcal{T})M(t,T,y)^2}{1+2C(T,\mathcal{T})\Sigma(t,T)}\right\}}{\sqrt{1 + 2C(T, \mathcal{T})\Sigma(t, T)}}. \tag{6.45}$$

By substituting the expression for M given in (6.15), and collecting equal powers of y, we obtain the following three identities

$$C(t, \mathcal{T}) - C(t, T) = \frac{C(T, \mathcal{T})E(t, T)^2}{1 + 2C(T, \mathcal{T})\Sigma(t, T)} \tag{6.46}$$

$$B(t, \mathcal{T}) - B(t, T) = \frac{B(T, \mathcal{T})E(t, T) - 2C(T, \mathcal{T})E(t, T)M_1(t, T)}{1 + 2C(T, \mathcal{T})\Sigma(t, T)} \tag{6.47}$$

$$A(t, \mathcal{T}) - A(t, T) = A(T, \mathcal{T}) - \tfrac{1}{2}\log\left(1 + 2C(T, \mathcal{T})\Sigma(t, T)\right)$$
$$+ \frac{\frac{1}{2}B(T,\mathcal{T})^2\Sigma(t,T) + B(T,\mathcal{T})M_1(t,T) - C(T,\mathcal{T})M_1(t,T)^2}{1+2C(T,\mathcal{T})\Sigma(t,T)} \tag{6.48}$$

with

$$M_1(t, T) = \int_t^T \sigma^2 E(s, T)B(s, T)\, ds. \tag{6.49}$$

These identities hold for all $t < T < \mathcal{T}$. For the special case $t = 0$, we can make the following observations. For all $T > 0$ we have $A(0, T) = \log D(0, T, 0)$, furthermore, from (6.28) we get that $M_1(0, T) = \alpha(T) - F(T)$. Using these relations combined with (6.47) and (6.48), we can express $B(T, \mathcal{T})$ and $A(T, \mathcal{T})$ as follows

$$B(T, \mathcal{T}) = \frac{1 + 2C(T, \mathcal{T})\Sigma(0, T)}{E(0, T)}\left(B(0, \mathcal{T}) - B(0, T)\right)$$
$$+ 2C(T, \mathcal{T})\left(\alpha(T) - F(T)\right) \tag{6.50}$$

$$A(T, \mathcal{T}) = \log\left(\frac{D(0, \mathcal{T}, 0)}{D(0, T, 0)}\right) + \tfrac{1}{2}\log\big(1 + 2C(T, \mathcal{T})\Sigma(0, T)\big)$$

$$-\frac{\tfrac{1}{2}B(T,\mathcal{T})^2\Sigma(0,T)+B(T,\mathcal{T})(\alpha(T)-F(T))-C(T,\mathcal{T})(\alpha(T)-F(T))^2}{1+2C(T,\mathcal{T})\Sigma(0,T)}. \quad (6.51)$$

Finally, we can calculate the values of $B(0, T)$ efficiently the following way. Substituting the definition of the forward rates into (6.25) we obtain

$$-\frac{\partial \log D(t, T, y)}{\partial T} = \Sigma(t, T) + \big(M(t, T, y) + \alpha(T)\big)^2. \quad (6.52)$$

Substituting the formulæ for $D(t, T, y)$ and $M(t, T, y)$ into this equation and collecting equal powers of y leads, again, to three identities. In this way, we get for B the identity

$$B_T(t, T) = 2E(t, T)\big(\alpha(T) - M_1(t, T)\big). \quad (6.53)$$

At $t = 0$ this identity simplifies to $B_T(0, T) = 2E(0, T)F(T)$, from which $B(0, T)$ can be calculated as

$$B(0, T) = 2 \int_0^T E(0, s)F(s)\, ds. \quad (6.54)$$

The values for $B(0, T)$ have to calculated only once for different T, and can then be stored.

7. An Empirical Comparison of One-Factor Models

This chapter is the final chapter in the first part of this book.[9] We have derived the general theory of valuing derivative securities, and we have shown how this theory can be used for valuing interest rate derivatives. We analysed in Chapters 5 and 6 a linear and a squared normal model which both have a rich analytical structure. However, only little attention has been devoted to the empirical validity of these models. In this chapter we address this problem.

Yield-curve models take the initial term-structure of interest rates (and hence bond- or swap-prices) as an input. To estimate the remaining parameters of the yield-curve models we use prices of actively traded US-dollar interest rate caps and floors, which are quoted on a broker screen. To make a comparison between yield-curve models, we have selected the following one-factor yield-curve models: a Hull-White model, a squared Gaussian model and a lognormal model. These models represent the most important solutions for modelling the yield-curve with spot rate models.

We have chosen such a specification for the models that they all have two unknown parameters determining the term-structure of volatilities. Using the observed cap and floor prices, the parameters can be estimated via non-linear least squares. To decide which of the three models provides the best empirical specification, we test the models against each other. However, the different yield-curve models are non-nested non-linear models. This implies that the standard t-tests cannot be used as specification tests. Hence, we use the P-test proposed by Davidson and MacKinnon (1993) to discriminate between the different models.

The results of the tests show that the lognormal model is the model that describes the observed cap and floor prices best. However, there is some evidence that a model with a distribution which is even more skewed to the right might provide a better fit.

The rest of this chapter is organised as follows. Section 7.1 describes the yield-curve models we want to consider. Section 7.2 describes the econometric

---

[9] The author is grateful to Andy Richardson at Intercapital Brokers for graciously providing the data on cap and floor prices. The author is also indebted to Frank de Jong for many comments and helpful suggestions. Some of the results presented in this chapter are also reported in Moraleda and Pelsser (2000).

approach to estimate and test the models. Section 7.3 describes the data. Section 7.4 presents and discusses the empirical results. Section 7.5 summarises and concludes.

7.1 Yield-Curve Models

The first model we want to consider is the Hull-White model (HW), which was analysed in Chapter 5. The spot interest rate r is described in this model by the stochastic differential equation

$$dr = \big(\theta(t) - ar\big)\, dt + \sigma\, dW. \tag{7.1}$$

The HW model has a rich analytical structure. Analytical expressions for prices of discount bonds can be obtained for this model. Prices for options on bonds can be expressed in terms of cumulative normal distribution functions.

A major disadvantage of the HW model, however, lies in the fact that negative interest rates can occur, due to the assumption that interest rates are normally distributed.

The second model in this study, is a model that precludes the possibility of negative interest rates, without losing the analytic tractability of the HW model. It is the one-factor squared Gaussian (SG) model, studied in Chapter 6

$$\begin{cases} du = \big(\theta(t) - au\big)\, dt + \sigma\, dW \\ r = u^2. \end{cases} \tag{7.2}$$

In this model the spot interest rate never becomes negative and has a non-central chi-square distribution with one degree of freedom.

In the SG model analytical expressions for prices of discount bonds can be obtained, also prices of options on discount bonds can be expressed in terms of cumulative normal distribution functions.

Another way to circumvent negative interest rates is to extend the square root model of Cox, Ingersoll and Ross (1985). Several extensions have been proposed, see e.g. Cox, Ingersoll and Ross (1985) and Hull and White (1990a), which allow the square root model to be fitted to the initial term-structure of interest rates but destroy the analytic properties of the square root model. Jamshidian (1995) proposes an extension of the square root model that preserves the analytical properties of the square root model, which he calls the simple square root model. Rogers (1995) shows that simple square root models are equivalent to multi-factor squared Gaussian models. Since we consider only one-factor models in this book, we have not included a simple square root model in this study.[10]

---

[10]We have investigated the exponential decay model proposed in the paper by Jamshidian (1995), which is an example of a simple square root model. This

The third model we consider is a lognormal model (LN). Lognormal yield-curve models have been developed by Black, Derman and Toy (1990) and later generalised by Black and Karasinski (1991). The version of the lognormal model we consider in this chapter is a restricted version of the Black-Karasinski model due to Hull and White (1994) where the spot interest rate r follows the process

$$d\log r = \left(\theta(t) - a\log r\right) dt + \sigma\, dW. \tag{7.3}$$

Here, the logarithm of the spot interest rate follows a Hull-White model, and is normally distributed. Hence, the spot interest rate itself has a lognormal distribution and the occurrence of negative interest rates is also impossible in the LN model. The largest disadvantage of the LN model is that no analytical formulæ are known for the prices of discount bonds and options on discount bonds. To calculate these prices we have to resort to numerical methods.

Consider the general normal model

$$\begin{cases} dy = -ay\, dt + \sigma\, dW \\ \ r = F\left(\alpha(t) + y\right). \end{cases} \tag{7.4}$$

In this general model, the underlying process y follows an Ornstein-Uhlenbeck process. The process y depends only on the volatility parameters a and σ, and is independent of the initial term-structure. A little algebra will reveal that for

$$\alpha(t) = e^{-at} \left(F^{-1}(r_0) + \int_0^t e^{as}\theta(s)\, ds \right) \tag{7.5}$$

with $F(x) \equiv x$ we obtain the HW model, with $F(x) \equiv x^2$ we obtain the SG model and with $F(x) \equiv e^x$ we obtain the LN model. Note that this is the general form of the transformation of variables used in Chapters 5 and 6. Since there is a one-to-one relationship between $\alpha(t)$ and $\theta(t)$, it is also possible to use $\alpha(t)$ directly to fit the models to the initial term-structure of interest rates, and obtain $\theta(t)$ via

$$\theta(t) = \frac{d}{dt}\alpha(t) + a\alpha(t), \tag{7.6}$$

which can be obtained from differentiating (7.5) with respect to t.

To calculate the prices of caps and floors, we can use for the HW and the SG models the analytic formulæ derived in Chapter 5 and 6, respectively. For the LN model, the prices for caps and floors have to be calculated numerically. In this study we have used the trinomial tree approach outlined in Hull and White (1994).

---

model cannot be fitted to all term-structures in our sample; also the parameters are very hard to estimate due to the fact that the objective function is very flat around the minimum. For the days the model can be fitted, it fits the observed cap and floor prices about as well as the squared Gaussian model.

7.2 Econometric Approach

The main focus of empirical work in the literature so far has been directed towards models with an endogenous term-structure of interest rates. Using the prices for actively traded (discount) bonds, several authors have estimated the parameters of several models, see e.g. Brown and Dybvig (1986), Longstaff and Schwartz (1992), Chan et al. (1992) and De Munnik and Schotman (1994).

One of the most interesting papers in this area is Chan et al. (1992). Using Treasury bill yield data they use the generalised method of moments (GMM) to estimate and test the model

$$dr = (\alpha - \beta r) \, dt + \sigma r^\gamma \, dW. \tag{7.7}$$

This general specification encompasses several of the endogenous term-structure models. For $\gamma = 0$ the Vasicek, and for $\gamma = \frac{1}{2}$ the Cox-Ingersoll-Ross model is obtained. For $\gamma = 1$ a kind of lognormal model is obtained. Although the stochastic process (7.7) is very general, it remains unclear whether it is well defined for $\gamma > 1$ (see Rogers (1995)). Their results have to be interpreted with some care. As explained earlier, models with an endogenous term-structure of interest rates cannot be fitted to the initial term-structure, which makes these models less attractive for pricing and managing interest rate derivatives.

For this reason we consider in this chapter yield-curve models, and we adopt a different approach. As was explained in the previous section, the function $\alpha(t)$ (and hence $\theta(t)$) can be determined from the initial term-structure. To determine the parameters a and σ for each model, we fit the models to observed cap and floor prices. Hull and White (1994) suggest that the parameters for their model can be estimated by minimizing the sum of squared differences between the observed cap/floor prices and the prices calculated by the model. However, in the data the observed prices on any given day can range from 1 basispoint to 999 basispoints, which is almost three orders of magnitude. Therefore we have chosen to fit the models to the observed prices in logarithmic terms, which is equivalent to minimizing the relative pricing errors of the models. A fit in logarithmic terms has the additional advantage that the heteroscedasticity of the errors is reduced. Consequently, the econometric model can be written as

$$y = g(X; a, \sigma) + \epsilon, \tag{7.8}$$

where y is the vector of the logarithms of observed cap and floor prices, and X is a matrix of explanatory variables, which in our case consists of the maturity, strike level and a binary variable indicating cap or floor for every contract.[11] The (vector valued) function g denotes the logarithm of the price

---

[11] The first three columns of Table 7.2 contain an example of X.

calculated in a model. Estimates $\hat{a}$ and $\hat{\sigma}$ for the parameters are obtained via non-linear least squares. If the errors are independently, identically distributed with zero mean and the function g is sufficiently differentiable, then the non-linear least squares estimates are consistent and have an asymptotical normal distribution (see, e.g. Judge et al. (1982)).

The intention of this study is to identify the model that fits the observed cap/floor prices best. Like the option models used for equity and foreign-exchange derivatives, yield-curve models are used as relative valuation tools by banks and institutions. Taking the market prices of actively traded derivatives as given, yield-curve models are calibrated to these prices and are then used to calculate prices and hedge ratios for more complicated (and less liquid) derivatives. For this reason it is important to find a parsimonious model that provides a good fit to observed market prices. Therefore, we do not impose the restriction that the parameters estimates $\hat{a}$ and $\hat{\sigma}$ of any of the models are constant over time. Consequently, we estimated and tested the models on a daily basis using the cap/floor prices observed on any given day. This way of treating the models is consistent with the way yield-curve models are used in practice to price and manage interest rate derivatives.

To determine which of the three yield-curve models describes the data best, we have to choose from the following three hypotheses

$$H_{\mathrm{HW}}: \quad y = g_{\mathrm{HW}}(X; a_{\mathrm{HW}}, \sigma_{\mathrm{HW}}) + \epsilon_{\mathrm{HW}} \tag{7.9}$$

$$H_{\mathrm{SG}}: \quad y = g_{\mathrm{SG}}(X; a_{\mathrm{SG}}, \sigma_{\mathrm{SG}}) + \epsilon_{\mathrm{SG}} \tag{7.10}$$

$$H_{\mathrm{LN}}: \quad y = g_{\mathrm{LN}}(X; a_{\mathrm{LN}}, \sigma_{\mathrm{LN}}) + \epsilon_{\mathrm{LN}}. \tag{7.11}$$

Two approaches can be adopted to determine the "best" model. One approach is based on goodness-of-fit criteria like R^2 or the standard error of the regression. These criteria are often easy to compute and intuitively appealing. However, goodness-of-fit criteria have several disadvantages. Using these criteria it is difficult to determine whether one model is "significantly" better than another model, or (even worse) whether all models describe the data badly. In other words, goodness-of-fit criteria do not take into consideration the losses associated with choosing an incorrect model (Judge et al. (1982), Chapter 22.4).

These problems can be avoided by formally testing models against each other. The three models under consideration lead to three hypotheses which are non-linear and non-nested. The testing of non-nested non-linear hypotheses is a hard problem, and no consensus exists which test procedure is to be preferred. For an overview of some test procedures that have been suggested see, for example, Fisher and McAleer (1981) and Mizon and Richard (1986). A class of tests which is intuitively appealing and easy to implement, are tests which are based on the principle of artificial nesting. Suppose we have two non-nested hypotheses,

$$H_1: \quad y = g(X; \gamma) + \epsilon_0 \tag{7.12}$$

$$H_2: \quad y = h(X; \delta) + \epsilon_1 \tag{7.13}$$

where γ and δ are the parameter-vectors, then we can create an artificial compound model

$$H_C: \quad y = (1-\beta)g(X;\gamma) + \beta h(X;\delta) + \epsilon. \tag{7.14}$$

The artificial parameter β has been introduced to nest H_1 and H_2 into H_C. The test is then based on the hypothesis $\beta = 0$ or $\beta = 1$. An additional advantage of the artificial compound model is that it can be used as a specification test. If β is significantly different from 0 and 1, it can provide an indication in which direction we have to search for a better model.

The artificial model H_C can, in general, not be estimated. If the nesting parameter β approaches either 1 or 0, the parameters γ or δ will not be identified. A solution to this problem (see e.g. Davidson and MacKinnon (1993)) is to replace the parameter vector of the alternative model by a consistent estimate of the parameter vector under the alternative hypothesis. If we take H_2 to be the alternative hypothesis, then we can replace δ by its non-linear least squares estimate $\hat{\delta}$, and we obtain the alternative model

$$y = (1-\beta)g(X;\gamma) + \beta\hat{h} + \epsilon, \tag{7.15}$$

where $\hat{h} = h(X;\hat{\delta})$. If H_1 and H_2 are really non-nested, then both β and γ are asymptotically identifiable and have an asymptotically normal distribution. The hypothesis H_1 can be tested, by testing the null hypothesis $\beta = 0$. To implement the test, one can use the t-statistic from the non-linear regression (7.15). This is known as the J-test.

If we use the Taylor approximation $g \approx \hat{g} + \hat{G}(\gamma - \hat{\gamma})$, where $\hat{G}$ is the matrix of partial derivatives of g with respect to γ evaluated at $\hat{\gamma}$, the non-linear regression (7.15) can be linearized to obtain the Gauss-Newton regression

$$y - \hat{g} = \hat{G}c + b(\hat{h} - \hat{g}) + \eta. \tag{7.16}$$

An alternative procedure to test H_1 is to use the t-statistic for $b = 0$ from the Gauss-Newton regression. This procedure is much simpler to implement and is asymptotically equivalent to the J-test under H_1 (see Davidson and MacKinnon (1993)) and is called the P-test. It is this test procedure we have implemented.

Since it is just as valid to test H_1 as H_2 against the composite hypothesis H_C, we can have four possible outcomes:

i　accept H_1 and reject H_2;
ii　reject H_1 and accept H_2;
iii　accept both hypotheses;
iv　reject both hypotheses.

In case *iii* the test indicates that both models are satisfactory, in case *iv* both models are rejected against the artificial compound model and we have to search for another model with more explanatory power. In this case, the

value of the nesting parameter can give an indication in which direction we
have to look for a better model.

7.3 Data

We have used data from two sources. Intercapital Brokers provided us with
data on cap and floor prices. For 1993 and 1994 they provided us with copies
from the Reuters page ICAV at 5:30pm (London time) each trading day.
This shows Intercapital's bid and offer quotes for actively traded US-dollar
3 months interest rate caps and floors, totalling 508 trading days. Note that
these prices are not closing prices, because US-dollar interest caps and floors
are traded continuously on an around the world basis. Each day Intercapital
Brokers show quotes for caps and floors with three different strike levels
(depending on the level of interest rates) and six different maturities (1,2,3,4,5
and 10 years), giving a total of 36 caps/floors. From the bid and offer quotes
mid-market prices were calculated to which the models were fitted.

The interest rates for every trading day were obtained from Datastream.
For every trading day in the sample we downloaded the overnight rate, the
1,3,6 and 12-months US-dollar money-market rates, and the 2,3,4,5,7 and 10-
year US-dollar swap-rates. From these observed rates, 11 continuously compounded zero-rates can be calculated. The complete zero-curve was obtained
by loglinear interpolation of the discount bond prices.

7.4 Empirical Results

As was explained above, the three models under consideration were estimated
and tested on a daily basis. In this section we summarise the results for these
daily regressions. First we present the parameter estimates and the standard
errors of the regressions of the individual models, then we show the results
of the pair wise P-tests we conducted.

Table 7.1, Figure 7.1 and Figure 7.2 show the results of the 508 daily
regressions. Table 7.1 contains summary statistics for the parameter estimates
of a and σ and the standard error of the regression σ_{LS} for each of the three
models. For the complete sample of 508 observations we report the mean and
standard deviation of the estimated parameters. Furthermore we show the
minimum, first quartile, median, third quartile and maximum value of the
estimated parameters over the sample.

Figure 7.1 shows σ_{LS} for all three models in graphical form. Because all
three models have two parameters, this measure is equivalent to R^2. However,
all three models have in general R^2s in excess of 0.99; hence the standard error
of the regression σ_{LS} shows the differences between the models more clearly.

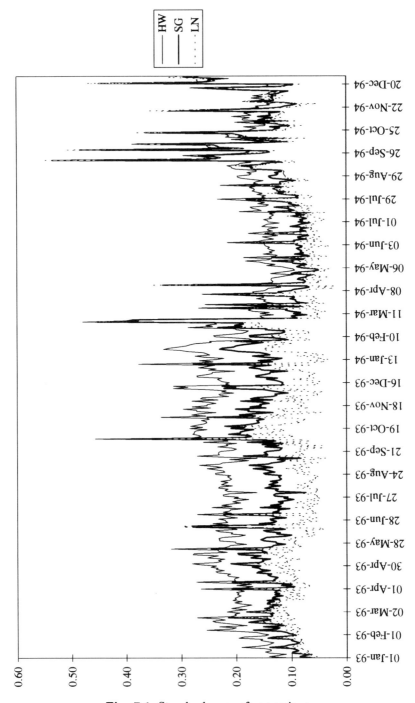

Fig. 7.1. Standard error of regressions

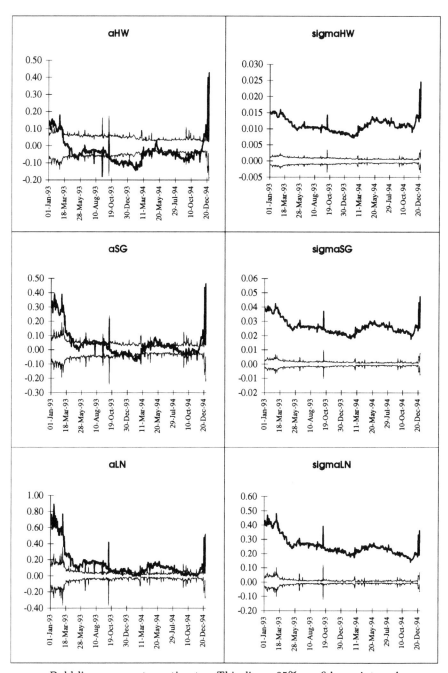

Bold line: parameter estimate   Thin lines: 95% confidence interval

Fig. 7.2. Parameter estimates

Table 7.1. Summary statistics for parameter estimates

		Mean	St.Dev.	Min.	Q1	Median	Q3	Max.
HW	$\hat{a}$	-0.031	0.076	-0.179	-0.072	-0.043	-0.023	0.426
	$\hat{\sigma}$	0.011	0.002	0.007	0.010	0.011	0.012	0.025
	σ_{LS}	0.206	0.058	0.093	0.161	0.204	0.237	0.531
SG	$\hat{a}$	0.045	0.099	-0.088	-0.016	0.024	0.056	0.463
	$\hat{\sigma}$	0.026	0.005	0.018	0.023	0.026	0.028	0.047
	σ_{LS}	0.142	0.059	0.050	0.133	0.133	0.159	0.535
LN	$\hat{a}$	0.149	0.168	0.000	0.051	0.103	0.160	0.888
	$\hat{\sigma}$	0.254	0.065	0.140	0.212	0.242	0.268	0.481
	σ_{LS}	0.108	0.068	0.023	0.067	0.091	0.122	0.548

Summary statistics computed over the complete sample of 508 days.
Q1 and Q3 denote first and third quartile.

Over the complete sample, the HW model fits the cap/floor prices worst with an average value for the standard error equal to 0.206. The SG model fits the data better with an average standard error of 0.142. The LN model fits the observed prices best, although towards the end of 1994 the difference between the models becomes less pronounced. The average standard error for the LN model is equal to 0.108. If the standard error of the regression (or R^2) is used as a model selection criterion, we would clearly choose the LN model. However, goodness-of-fit criteria are generally considered to be a poor model selection criterion.

The individual parameter estimates for the three models are shown in graphical form in Figure 7.2. For every parameter we plotted the estimates and $+2$ and -2 times the asymptotic standard error of the estimates, which gives an indication whether the estimate is significantly different from zero. For all three models, the parameter σ is significantly different from zero, often more than 10 standard deviations away from zero. For the mean-reversion parameter a the picture is entirely different. On the basis of economic theory, there are compelling arguments for the mean-reversion of interest rates. This should be reflected by the fact that the parameter a is significantly positive. As can be seen from the graphs, a_{HW} is significantly negative for prolonged periods in our sample. The SG model has a mean-reversion parameter which, except for the first few months, is not significantly different from zero. Only the LN model has a mean-reversion parameter that is significantly positive for the larger part of the sample.[12]

The fact that a_{HW} and a_{SG} do not have the correct sign, can be seen as an indication that the HW and the SG model do not describe the empirical steady-state distribution of interest rates adequately. If the empirical steady-state distribution of interest rates has a relatively fat tail on the right, the only

---

[12]Although the LN model is well-defined for $a_{LN} < 0$, we cannot build a trinomial tree in that case. Therefore, we estimated the LN model with the restriction $a_{LN} > 0$. However, the restriction was binding for only 21 days in the sample.

Table 7.2. Observed and fitted cap/floor prices on 04-Jan-94

Mat.	C/F	Strike(%)	Bid–Ask	HW	SG	LN
				Prices (in bp)		
1yr	C	3.25	53–55	46	46	45
	C	3.50	37–39	32	31	31
	C	3.75	24–27	21	20	20
	F	3.75	10–12	16	15	15
	F	3.50	03–06	8	7	7
	F	3.25	02–04	4	3	2
2yr	C	5.00	45–49	42	44	47
	C	5.50	26–30	21	24	28
	C	6.00	15–18	10	12	16
	F	4.50	72–76	73	74	76
	F	4.00	30–34	34	34	34
	F	3.50	07–10	11	10	9
3yr	C	5.00	138–146	129	134	139
	C	5.50	96–103	82	88	96
	C	6.00	66–73	48	56	66
	F	4.50	90–97	89	91	93
	F	4.00	38–44	43	42	42
	F	3.50	10–13	16	14	12
4yr	C	5.00	270–284	255	262	268
	C	5.50	203–217	178	188	199
	C	6.00	152–165	120	133	147
	F	4.50	104–116	105	107	109
	F	4.00	46–54	52	51	50
	F	3.50	13–19	21	18	15
5yr	C	6.50	196–214	154	174	195
	C	7.00	152–168	107	127	150
	C	7.50	120–134	72	92	115
	F	5.50	336–356	309	321	333
	F	5.00	216–236	204	211	216
	F	4.50	124–138	122	123	124
10yr	C	6.50	636–686	712	707	697
	C	7.00	524–574	580	584	582
	C	7.50	432–482	471	482	487
	F	5.50	480–540	575	548	520
	F	5.00	310–360	419	381	345
	F	4.50	176–216	294	247	207
			$\hat{a}$:	−0.118	−0.060	0.035
			$\hat{\sigma}$:	0.008	0.020	0.206

way the HW and SG models can mimic this is by increasing the long-term variance, which can be achieved by setting the mean-reversion parameter to a small or negative value.

To get an indication of the fit of the three models, we have reported in Table 7.2 the observed and fitted cap and floor prices on January 4, 1994. This day is observation day number 254, and is in the middle of the sam-

Table 7.3. Results of pair-wise P-tests

sign. lvl	acc. HW/rej. SG		rej. HW/acc. SG		acc. HW&SG		reject HW&SG	
10%	0	0%	55	11%	27	5%	426	84%
5%	3	1%	43	8%	14	3%	448	88%
1%	4	1%	40	8%	9	2%	455	90%
	acc. LN/rej. HW		rej. LN/acc. HW		acc. LN&HW		reject LN&HW	
10%	393	77%	10	2%	20	4%	85	17%
5%	335	66%	9	2%	8	2%	156	31%
1%	297	58%	10	2%	4	1%	197	39%
	acc. LN/rej. SG		rej. LN/acc. SG		acc. LN&SG		reject LN&SG	
10%	342	67%	16	3%	73	14%	77	15%
5%	305	60%	20	4%	45	9%	138	27%
1%	273	54%	26	5%	29	6%	180	35%

Column entries are: # of days; % of sample (508 days)

ple. The first thing to notice is that for longer maturities, the differences between the models becomes more pronounced. For short maturities there is relatively little uncertainty in the interest rates. This means that for short maturities the distribution of interest rates is relatively spiked and can be well approximated by a normal distribution. For longer maturities the uncertainty over future interest rates becomes larger, and the differences between the normal, chi-square and lognormal distribution is prominently reflected in the prices. For longer maturities it is clear that the fit of the LN model is much better than the fit of the HW and the SG model. The differences in skewness of the distributions implied by the three models are reflected by the fact that higher values are assigned to out-of-the-money caps and lower values to out-of-the-money floors as the distributions become more skewed to the right.

To cast the model selection problem into a more formal setting, we have implemented three sets of pair-wise P-tests among the three models. The results of these tests are reported in Figure 7.3 and Table 7.3. Figure 7.3 shows the estimated nesting parameter b from the P-test regressions, and $+2$ and -2 times the asymptotic standard error of the estimate. As explained in Section 7.2, we can consider each of the two models in a P-test as the "null-hypothesis" which can lead to four outcomes of the P-test. Table 7.3 summarises how many times each of the outcomes occur at different significance levels for the three pair wise tests conducted.

Both the HW and the SG model are overwhelmingly rejected against the data. The estimations of the nesting parameters indicate consistently that a superior model is obtained by taking two times the SG model minus one time the HW model. This may seem a nonsensical model at first sight. However, by making this combination of models the artificial compound model tries to create a model which is more skewed to the right than the SG model. The

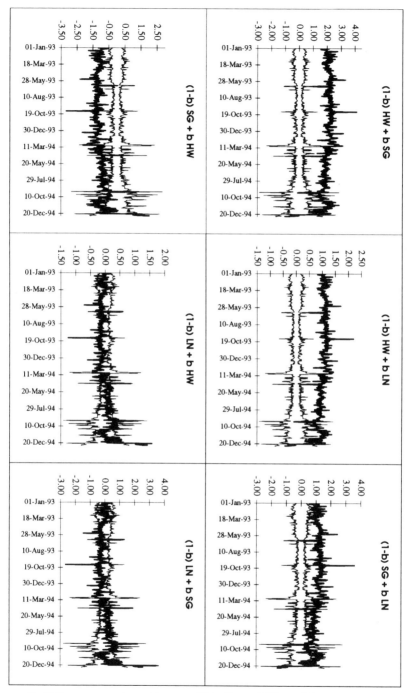

Bold line: parameter estimate   Thin lines: 95% confidence interval

Fig. 7.3. Pair-wise *P*-test regressions

results of this P-test point clearly in the direction of a lognormal model. If we test the LN model against the other two models, this is confirmed. In 60 to 70% of the sample the LN model is accepted as the correct model; however for some 30% of the sample all models are rejected. These results indicate that the LN model does a fairly good job in describing the observed cap and floor prices and is certainly the best model of the three models. From the estimated values of the nesting parameters we see that (especially for the first half of the sample) there is still some evidence that a distribution more skewed to the right than the lognormal distribution would provide a still better fit. However, towards the end of the sample, the dataset becomes somewhat less informative and the tests cannot clearly distinguish between the models.

It is interesting to note that these results are consistent with the findings of Chan et al. (1992), although they use a very different approach. The value for γ Chan et al. (1992) find after estimating (7.7) is approximately 1.5 with a standard error of 0.25, indicating that a lognormal model cannot be rejected but also indicating that a more skewed distribution would be appropriate. However, it remains unclear whether the stochastic process (7.7) for r is well defined for $\gamma > 1$, see Rogers (1995).

7.5 Conclusions

In this chapter we have empirically compared three one-factor spot rate models. We have considered a normal, a squared Gaussian and a lognormal model, which represent the three main types of spot rate models that have been proposed in the literature. To compare the models we have tested how well the models fit observed US-dollar interest rate caps and floors over a two year sample period. We find that, of the models considered, the lognormal model provides the best fit to the observed cap and floor prices. However, lognormal spot rate models have very little analytical properties and all prices must be calculated using numerical methods. In the next part of this book we will turn our attention to market rate models, which assume lognormal distributions for the traded interest rates like LIBOR or swap rates.

Part II

Market Rate Models

8. LIBOR and Swap Market Models

In the first part of this book, we have investigated various spot rate models which use the spot interest rate as a basis for modelling. The mathematically convenient choice for the spot interest rate leads to models which are particularly tractable. However, since these models are set up in terms of a mathematically convenient rate that does not exist in practice, valuation formulæ for real-world instruments like caps, floors and swaptions tend to be fairly complicated. To calibrate the spot rate models to the prices of these instruments we need complicated numerical procedures and the results are not always satisfactory.

This shortcoming has inspired the second type of approach, which takes real market interest rates, like LIBOR or swap rates, as a basis for modelling. These models are called *market rate models* and will be the focus of attention in the second part of this book. Market rate models tend to be more complicated in their setup, but the big advantage is that market standard pricing formulæ for the standard instruments can be reproduced with these models. Hence, by construction, these models can be made to fit the market prices perfectly.

In this chapter we discuss a class of models known as *market models*. These models postulate a geometric Brownian motion for the market rates under consideration, such that the Black (1976) formula is recovered for the price of a European option on the market rate. The Black formula is the market standard for calculating prices of European-style interest rate options. The LIBOR market models were introduced by Miltersen, Sandmann and Sondermann (1997) and Brace, Gatarek and Musiela (1997). The swap market model was introduced by Jamshidian (1998).

The rest of this chapter is organised as follows. First, we show in Sections 8.1 and 8.2 how the LIBOR market model and the swap market model can be set up in an arbitrage-free economy. We also show how the Black formula can be recovered to price caps and floors (in the LIBOR market model) and swaptions (in the swap market model). To calculate prices for other derivatives, we have to use Monte Carlo simulation. How to implement Monte Carlo simulation for LIBOR market models is explained in Section 8.3. Finally, we show in Section 8.4 how to implement Monte Carlo simulation for swap market models. Both these sections have worked examples for exotic interest rate options that are actually traded in the market.

8.1 LIBOR Market Models

The marketed assets in an interest rate economy are the discount bonds with different maturities. Let $D(t, T)$ denote the value at time t of a discount bond which pays 1 at maturity T.

If you put your money in a money-market account for a given period, the interest earned over this period is quoted as a LIBOR rate. At the end of a period of length ΔT, one receives an interest equal to αL, where L denotes the LIBOR rate and $\alpha = \Delta T$ denotes the *accrual factor* or *daycount fraction*.[13] Hence, we obtain the relation $1 \equiv (1 + \alpha L)D(0, \Delta T)$, which states that the present value today of the notional plus the interest earned at the end of a ΔT period is equal to the notional.

A *forward* LIBOR *rate* $L_{TS}(t)$ is the interest rate one can contract for at time t to put money in a money-market account for the time period $[T, S]$. We define the forward LIBOR rate via the relation

$$D(t, T) = \bigl(1 + \alpha_{TS} L_{TS}(t)\bigr) D(t, S), \tag{8.1}$$

where α_{TS} denotes the daycount fraction for the period $[T, S]$. Solving for L yields

$$L_{TS}(t) = \frac{1}{\alpha_{TS}} \left(\frac{D(t, T) - D(t, S)}{D(t, S)} \right). \tag{8.2}$$

The time T in known as the *maturity* of the forward LIBOR rate and $(S - T)$ is called the *tenor*.

At time T the forward LIBOR rate $L_{TS}(T)$ is *fixed* or *set* and is then called a *spot* LIBOR *rate*. Note, that the spot LIBOR rate is fixed at the beginning of the period at T, but is paid at the end of the period at S.

8.1.1 LIBOR Process

In most markets, only forward LIBOR rates of one specific tenor ΔT are actively traded, which is usually 3 months (e.g., for USD) or 6 months (e.g., for EURO). Therefore, we assume there are N forward LIBOR rates with this specific tenor, which we denote by $L_i(t) = L_{T_i T_{i+1}}(t)$ and $T_i = i\Delta T$ for

---

[13]Note that in real markets α is not exactly equal to ΔT, but is calculated according to a specific algorithm for a given market known as the *daycount convention*. There are about a dozen different conventions used, hence one should check carefully to use the correct convention for the market under consideration.

$i = 1, \ldots, N$, with daycount fractions $\alpha_i = \alpha_{T_i, T_{i+1}}$.[14] For this set of LIBOR rates, we denote the associated discount factors by $D_i(t) = D(t, T_i)$.

Let us concentrate on $L_i(t)$. The process

$$\alpha_i L_i(t) = \frac{D_i(t) - D_{i+1}(t)}{D_{i+1}(t)} \tag{8.3}$$

is a ratio of marketed assets. Hence, if we take the discount bond D_{i+1} as a numeraire, then (given that the economy is arbitrage-free) under the martingale measure $\mathbb{Q}^{i+1}$ associated with the numeraire D_{i+1} the process $\alpha_i L_i$ will be a martingale. As α_i is a constant, the process $L_i(t)$ must also be a martingale under $\mathbb{Q}^{i+1}$.

The LIBOR market model makes the assumption that L_i is given by the stochastic differential equation

$$dL_i(t) = \sigma_i(t) L_i(t)\, dW^{i+1}, \tag{8.4}$$

where W^{i+1} denotes a Brownian motion under $\mathbb{Q}^{i+1}$. The solution to equation (8.4) can be expressed as

$$L_i(t) = L_i(0) \exp\left\{ -\frac{1}{2} \int_0^t \sigma_i(s)^2\, ds + \int_0^t \sigma_i(s)\, dW^{i+1}(s) \right\}. \tag{8.5}$$

If $\sigma_i(t)$ is a deterministic function, then L_i has a lognormal probability distribution under $\mathbb{Q}^{i+1}$, where the variance of $\log L_i(t)$ is equal to $\int_0^t \sigma_i(s)^2\, ds$.

8.1.2 Caplet Price

As explained in Chapter 5 (Sect. 5.5), a cap contract is a portfolio of options on LIBOR rates called caplets. A caplet is an insurance against high interest rates. Suppose we have an insurance that protects us from the LIBOR rates L_i fixing above a level K. The payoff we receive from the caplet $\mathbf{C}_i$ at time T_{i+1} is equal to

$$\mathbf{C}_i(T_{i+1}) = \alpha_i \max\{L_i(T_i) - K, 0\}. \tag{8.6}$$

This payoff exactly matches the difference between the insured payment $\alpha_i K$ and the LIBOR payment $\alpha_i L_i(T_i)$ that has to be made at time T_{i+1}. A caplet is therefore a call option on a LIBOR rate.

If we choose D_{i+1} as a numeraire and work under the associated measure $\mathbb{Q}^{i+1}$ we know that $\mathbf{C}_i / D_{i+1}$ is a martingale and we obtain

$$\frac{\mathbf{C}_i(0)}{D_{i+1}(0)} = \mathbb{E}^{i+1}\left(\frac{\mathbf{C}_i(T_{i+1})}{D_{i+1}(T_{i+1})} \right) = \alpha_i \mathbb{E}^{i+1}\left(\max\{L_i(T_i) - K, 0\} \right), \tag{8.7}$$

---

[14]Again, in real markets the dates T_i are not spaced exactly ΔT apart, but are determined with a specific algorithm for a given market known as the *date-roll convention*.

where $\mathbb{E}^{i+1}$ denotes expectation with respect to the measure $\mathbb{Q}^{i+1}$.

Given the lognormal assumption on the process of L_i, we can calculate the expectation explicitly as

$$\mathbf{C}_i(0) = \alpha_i D_{i+1}(0)\big(L_i(0)N(d_1) - KN(d_2)\big) \tag{8.8}$$

where $d_1 = \big(\log(L_i(0)/K) + \frac{1}{2}\Sigma_i^2\big)/\Sigma_i$, $d_2 = d_1 - \Sigma_i$ and $\Sigma_i^2 = \int_0^{T_i} \sigma_i(s)^2\, ds$ is the variance of $\log L_i(T_i)$.

Using the same arguments, it follows that the price of a floorlet $\mathbf{F}_i$ is given by

$$\mathbf{F}_i(0) = \alpha_i D_{i+1}(0)\big(KN(-d_2) - L_i(0)N(-d_1)\big). \tag{8.9}$$

We see that in the LIBOR market model the valuation formula for caplets and floorlets is the Black (1976) formula used in the market to price these instruments. To calibrate the LIBOR market model, we only have to ensure that $\Sigma_i^2 = \bar\sigma_i^2 T_i$, where $\bar\sigma_i$ is the implied caplet volatility quoted in the market.

Note that the LIBOR market model is set up by specifying the dynamics of each forward LIBOR rate L_i with respect to the probability measure $\mathbb{Q}^{i+1}$. Hence, a different probability measure is used for each LIBOR rate. To price European options like caplets this is not a problem, and indeed very convenient as each LIBOR rate is a martingale in its "own" probability measure. However, if we want to determine the price for more complicated derivatives, we need to model the behaviour of all LIBOR rates simultaneously under a single measure.

8.1.3 Terminal Measure

To bring all the LIBOR rates under the same measure, we first consider the change of measure $d\mathbb{Q}^i/d\mathbb{Q}^{i+1}$. By repeated application of this change of measure for different i, we can bring all forward LIBOR processes under the same measure.

By the Change of Numeraire Theorem we know that the Radon-Nikodym derivative $\rho(t)$ is given by

$$\frac{d\mathbb{Q}^i}{d\mathbb{Q}^{i+1}} = \rho(t) = \frac{D_i(t)/D_i(0)}{D_{i+1}(t)/D_{i+1}(0)} = \frac{D_{i+1}(0)}{D_i(0)}\big(1 + \alpha_i L_i(t)\big), \tag{8.10}$$

where we applied (8.2) in the last equality.

To apply Girsanov's Theorem we need to find the process $\kappa(t)$ such that

$$\rho(t) = \exp\left\{\int_0^t \kappa(s)\, dW^{i+1}(s) - \frac{1}{2}\int_0^t \kappa(s)^2\, ds\right\}. \tag{8.11}$$

An application of Itô's Lemma shows that $d\rho(t) = \rho(t)\kappa(t)\, dW^{i+1}(t)$. Hence, $\kappa(t)$ is the "volatility" of the Radon-Nikodym derivative $\rho(t)$. Using (8.10) and (8.4) we find

$$dp(t) = \frac{\alpha_i \sigma_i(t) L_i(t)}{1 + \alpha(i) L_i(t)} \rho(t)\, dW^{i+1} \tag{8.12}$$

and we can identify $\kappa(t) = \alpha_i \sigma_i(t) L_i(t) / \big(1 + \alpha_i L_i(t)\big)$. Girsanov's Theorem now gives the relation

$$dW^i = dW^{i+1} - \frac{\alpha_i \sigma_i(t) L_i(t)}{1 + \alpha_i L_i(t)} dt, \tag{8.13}$$

where W^i and W^{i+1} are Brownian motions under the measures $\mathbb{Q}^i$ and $\mathbb{Q}^{i+1}$ respectively.

If we consider the N LIBOR rates in our economy, we can take the terminal discount bond D_{N+1} as the numeraire and work under the measure $\mathbb{Q}^{N+1}$ which is called the *terminal measure*. Under the terminal measure, the terminal LIBOR rate L_N is a martingale.

The process for L_{N-1} can be derived using Girsanov's Theorem. Applying (8.13) for $i = N$ we obtain that L_{N-1} follows under the measure $\mathbb{Q}^{N+1}$ the process

$$
\begin{aligned}
dL_{N-1}(t) &= \sigma_{N-1}(t) L_{N-1}(t) \left(dW^{N+1} - \frac{\alpha_N \sigma_N(t) L_N(t)}{1 + \alpha_N L_N(t)} dt \right) \\
&= -\frac{\alpha_N \sigma_N(t) L_N(t)}{1 + \alpha_N L_N(t)} \sigma_{N-1}(t) L_{N-1}(t)\, dt \\
&\qquad + \sigma_{N-1}(t) L_{N-1}(t)\, dW^{N+1}.
\end{aligned}
\tag{8.14}
$$

If we use (8.13) repeatedly, we can derive that L_i, in general, follows under the terminal measure the process

$$dL_i(t) = -\sum_{k=i+1}^{N} \frac{\alpha_k \sigma_k(t) L_k(t)}{1 + \alpha_k L_k(t)} \sigma_i(t) L_i(t)\, dt + \sigma_i(t) L_i(t)\, dW^{N+1}, \tag{8.15}$$

for all $1 \leq i \leq N$.

We see that, apart from the terminal LIBOR rate L_N, all LIBOR rates are no longer martingales under the terminal measure, but have a drift term that depends on the forward LIBOR rates with longer maturities. As (8.15) is fairly complicated we cannot solve the stochastic differential equations analytically, but we have to use numerical methods like Monte Carlo simulation to solve it. We will show how to do this in Section 8.3.

8.2 Swap Market Models

Swap market models describe the evolution of forward swap rates. We will first explain what swaps are, and how forward swap rates are defined. Then we will derive the swap market model.

8.2.1 Interest Rate Swaps

An *interest rate swap* is a contract where two parties agree to exchange a set of floating interest rate payments for a set of fixed interest rate payments. The set of floating interest rate payments is based on LIBOR rates and is called the *floating leg*. The set of fixed payments is called the *fixed leg*. The naming convention for swaps is based on the fixed side. In a *payer swap* you pay the fixed side, in a *receiver swap* you receive the fixed side.

Given a set of payment dates T_i where the payments are exchanged, we can determine the value of a swap as follows. A floating interest payment made at time T_{i+1} is based on the LIBOR fixing $\alpha_i L_i(T_i)$. Hence, the present value $V_i^{\text{flo}}(t)$ of this payment is

$$
\begin{aligned}
V_i^{\text{flo}}(t) &= D_{i+1}(t)\mathbb{E}^{i+1}\left(\alpha_i L_i(T_i)\right) = D_{i+1}(t)\alpha_i L_i(t) \\
&= D_i(t) - D_{i+1}(t),
\end{aligned}
\tag{8.16}
$$

where we have used the fact that L_i is a martingale under the measure $\mathbb{Q}^{i+1}$.

Given a fixed rate K, the fixed payment made at time T_{i+1} is equal to $\alpha_i K$. Hence, the present value $V_i^{\text{fix}}(t)$ of this payment is given by

$$
V_i^{\text{fix}}(t) = D_{i+1}(t)\alpha_i K.
\tag{8.17}
$$

In a swap multiple payments are exchanged. Let $V_{n,N}^{\text{pswap}}(t)$ denote the value of a payer swap at time t that starts at T_n and ends at T_N. At the start date T_n the first LIBOR rate is fixed. Actual payments are exchanged at dates $T_{n+1}, \ldots, T_N$. The *swap tenor* is defined as $T_N - T_n$.

Given (8.16) and (8.17) we can determine the present value of the payer swap as

$$
\begin{aligned}
V_{n,N}^{\text{pswap}}(t) &= \sum_{i=n}^{N-1} V_i^{\text{flo}}(t) - \sum_{i=n}^{N-1} V_i^{\text{fix}}(t) \\
&= \left(D_n(t) - D_N(t)\right) - K \sum_{i=n}^{N-1} \alpha_i D_{i+1}(t).
\end{aligned}
\tag{8.18}
$$

The value of a receiver swap is given by

$$
V_{n,N}^{\text{rswap}}(t) = K \sum_{i=n}^{N-1} \alpha_i D_{i+1}(t) - \left(D_n(t) - D_N(t)\right).
\tag{8.19}
$$

In the market, swaps are not quoted as prices for different fixed rates K, but only the fixed rate K is quoted for each swap such that the present value of the swap is equal to zero. This particular rate is called the *par swap rate*. We denote the par swap rate for the $[T_n, T_N]$ swap with $y_{n,N}$. Solving (8.18) (or (8.19)) for $K = y_{n,N}$ such that $V_{n,N}^{\text{swap}}(t) = 0$ yields

$$y_{n,N}(t) = \frac{D_n(t) - D_N(t)}{\sum\limits_{i=n+1}^{N} \alpha_{i-1} D_i(t)}. \tag{8.20}$$

The term in the denominator is called the *accrual factor* or *present value of a basispoint* or PVBP. We denote the PVBP by

$$P_{n+1,N}(t) = \sum_{i=n+1}^{N} \alpha_{i-1} D_i(t). \tag{8.21}$$

With the PVBP, the forward par swap rate can be expressed as

$$y_{n,N}(t) = \frac{D_n(t) - D_N(t)}{P_{n+1,N}(t)}. \tag{8.22}$$

Note that a one-period swap rate $y_{i,i+1}$ is equal to the LIBOR rate L_i.

Given the par swap rate $y_{n,N}(t)$ we can calculate the value of a swap with a different fixed rate K as

$$V_{n,N}^{\text{pswap}}(t) = (y_{n,N}(t) - K) P_{n+1,N}(t) \tag{8.23}$$

$$V_{n,N}^{\text{rswap}}(t) = (K - y_{n,N}(t)) P_{n+1,N}(t). \tag{8.24}$$

8.2.2 Swaption Price

The PVBP is a portfolio of traded assets and has strictly positive value. Therefore, a PVBP can be used as a numeraire. If we use the PVBP $P_{n+1,N}(t)$ as a numeraire, then under the measure $\mathbb{Q}^{n+1,N}$ associated with the numeraire $P_{n+1,N}$ all $P_{n+1,N}$ rebased values must be martingales in an arbitrage-free economy. In particular, the par swap rate $y_{n,N}$ must be a martingale under $\mathbb{Q}^{n+1,N}$.

The swap market model makes the assumption that $y_{n,N}$ is a lognormal martingale under $\mathbb{Q}^{n+1,N}$. We therefore assume that $y_{n,N}(t)$ follows

$$dy_{n,N}(t) = \sigma_{n,N}(t) y_{n,N}(t) \, dW^{n+1,N}, \tag{8.25}$$

where $W^{n+1,N}$ is a Brownian motion under $\mathbb{Q}^{n+1,N}$ and $\sigma_{n,N}(t)$ is a deterministic function.

A *swaption* (short for swap option) gives the right to enter at time T_n into a swap with fixed rate K. A *receiver swaption* gives the right to enter into a receiver swap, a *payer swaption* gives the right to enter into a payer swap. Swaptions are often denoted as $T_n \times (T_N - T_n)$, where T_n is the option expiry date (and also the start date of the underlying swap) and $(T_N - T_n)$ is the tenor of the underlying swap.

For a payer swaption, it is of course only beneficial to enter the swap if the present value of the swap is positive. Hence, the value $\mathbf{PS}_{n,N}(T_n)$ of a payer swaption at time T_n is

$$\mathbf{PS}_{n,N}(T_n) = \max\{V_{n,N}^{\mathrm{pswap}}(T_n), 0\}. \tag{8.26}$$

If we use $P_{n+1,N}$ as a numeraire, we can calculate the value of the payer swaption under the measure $\mathbb{Q}^{n+1,N}$ as

$$\frac{\mathbf{PS}_{n,N}(0)}{P_{n+1,N}(0)} = \mathbb{E}^{n+1,N}\left(\frac{\max\{V_{n,N}^{\mathrm{pswap}}(T_n), 0\}}{P_{n+1,N}(T_n)}\right)$$

$$= \mathbb{E}^{n+1,N}\left(\max\{y_{n,N}(T_n) - K, 0\}\right). \tag{8.27}$$

Given the lognormal assumption on the process of $y_{n,N}$, we can calculate the expectation explicitly as

$$\mathbf{PS}_{n,N}(0) = P_{n+1,N}(0)\left(y_{n,N}(0)N(d_1) - KN(d_2)\right) \tag{8.28}$$

where $d_1 = \left(\log(y_{n,N}(0)/K) + \frac{1}{2}\Sigma_{n,N}^2\right)/\Sigma_{n,N}$, $d_2 = d_1 - \Sigma_{n,N}$ and $\Sigma_{n,N}^2 = \int_0^{T_n} \sigma_{n,N}(s)^2\, ds$ is the variance of $\log y_{n,N}(T_n)$.

Using the same arguments, it follows that the price of a receiver swaption $\mathbf{RS}_{n,N}$ is given by

$$\mathbf{RS}_{n,N}(0) = P_{n+1,N}(0)\left(KN(-d_2) - y_{n,N}(0)N(-d_1)\right). \tag{8.29}$$

We see that in the swap market model the valuation formula for swaptions is the Black (1976) formula used in the market to price these instruments. To calibrate the swap market model, we only have to ensure that $\Sigma_{n,N}^2 = \bar{\sigma}_{n,N}^2 T_n$, where $\bar{\sigma}_{n,N}$ is the implied swaption volatility quoted in the market.

Note, that the swap market model is set up by specifying the dynamics of each par swap rate $y_{n,N}$ with respect to the probability measure $\mathbb{Q}^{n+1,N}$. If we want to determine the price for more complicated derivatives, we need to model the behaviour of all swap rates simultaneously under a single measure.

However, we cannot specify a distribution for *all* swap rates simultaneously. If we consider an economy with $N + 1$ payment dates $T_1, \ldots, T_{N+1}$ we can model the $N + 1$ discount factors associated with these dates. Given that we choose one of these discount bonds as the numeraire, we have N degrees of freedom left to model. Therefore, we can only model N par swap rates, just like in the LIBOR market model where we have N LIBOR rates. All other swap rates (including the LIBOR rates) are then determined by the N "reference" swap rates. Only the reference swap rates can be modelled as lognormal in their own measure, the probability distributions of the other swap rates is then determined by the distribution of the reference rates and will, in general, not be lognormal. This implies in particular that the LIBOR market model and the swap market model are inconsistent with each other.

Let us consider three examples of choosing the N swap rates:

- The set of swap rates $y_{n,N+1}$ for $n = 1, \ldots, N$. This set of swap rates is suited for modelling products where the underlying swaps share the same final payment date. The terminal measure $\mathbb{Q}^{N+1}$ is a convenient measure to work in. We derive the swap market model under this measure in Section 8.2.3.

- The set of swap rates $y_{1,n}$ for $n = 2, \ldots, N + 1$. This choice is suited for modelling products where the underlying swaps share the same start date (in this case T_1). The T_1-forward measure $\mathbb{Q}^1$ is a convenient measure to work in. We derive the swap market model under this measure in Section 8.2.4.

- The set of swap rates $y_{n,n+1}$ for $n = 1, \ldots, N$. This is, of course, the LIBOR market model of Section 8.1.

8.2.3 Terminal Measure

In this subsection, we consider an economy with N par swap rates $y_{n,N+1}$ for $n = 1, \ldots, N$. These swaps have payments on the dates $T_1, \ldots, T_{N+1}$. To simplify the notation, we denote in this subsection $y_{n,N+1}$ by y_n.

For each swap rate y_n we assume that under the measure $\mathbb{Q}^{n+1,N+1}$ its process is given by (8.25). We want to bring all these processes under the terminal measure $\mathbb{Q}^{N+1}$.

First we observe that y_N is the LIBOR rate L_N which is a martingale under $\mathbb{Q}^{N+1}$. Hence, for y_N we are done.

Next we consider the penultimate swap rate y_{N-1}. The appropriate change of measure is described by $d\mathbb{Q}^{N+1}/d\mathbb{Q}^{N,N+1}$. However, to apply Girsanov's Theorem directly would lead to a quite complicated calculation. Therefore, we derive the appropriate drift term for y_{N-1} using a different method.

From the definition (8.22) of a forward swap rate we can derive the following recursive relation between the PVBP's

$$D_n - D_{N+1} = y_n P_{n+1,N+1}$$

$$\Downarrow$$

$$\alpha_{n-1} D_n = \alpha_{n-1} D_{N+1} + \alpha_{n-1} y_n P_{n+1,N+1}$$

$$\Downarrow$$

$$P_{n,N+1} - P_{n+1,N+1} = \alpha_{n-1} D_{N+1} + \alpha_{n-1} y_n P_{n+1,N+1}$$

$$\Downarrow$$

$$P_{n,N+1} = \alpha_{n-1} D_{N+1} + P_{n+1,N+1}(1 + \alpha_{n-1} y_n). \tag{8.30}$$

Substituting for $n = N - 1$ and $n = N$ yields

$$\frac{P_{N-1,N+1}}{D_{N+1}} = \alpha_{N-2} + \alpha_{N-1}(1 + \alpha_{N-2} y_{N-1})$$

$$+ \alpha_N (1 + \alpha_{N-1} y_N)(1 + \alpha_{N-2} y_{N-1}). \tag{8.31}$$

Under the measure $\mathbb{Q}^{N+1}$ the left-hand side of (8.31) is a martingale. Hence, the right-hand side of (8.31) must be a martingale as well. Let us denote

$P_{N-1,N+1}/D_{N+1}$ by P' We can apply Itô's Lemma to P' and we obtain the following expression for the drift term of P' under the measure $\mathbb{Q}^{N+1}$

$$\left(\mu_N \frac{\partial P'}{\partial y_N} + \mu_{N-1} \frac{\partial P'}{\partial y_{N-1}} \right.$$
$$+ \tfrac{1}{2}\sigma_N^2 y_N^2 \frac{\partial^2 P'}{\partial y_N^2} + \tfrac{1}{2}\sigma_{N-1}^2 y_{N-1}^2 \frac{\partial^2 P'}{\partial y_{N-1}^2} \tag{8.32}$$
$$\left. + \rho_{N,N-1}\sigma_N y_N \sigma_{N-1} y_{N-1} \frac{\partial^2 P'}{\partial y_N \partial y_{N-1}} \right) dt$$

where μ_n denotes the drift term of y_n, σ_n denotes $\sigma_{n,N+1}$ and $\rho_{N,N-1}$ denotes the correlation between y_N and y_{N-1}. In a one-factor model this correlation is equal to 1, but we have included the correlation here to show how multi-factor models can be derived.

Since P' is a martingale, the expression (8.32) must be equal to zero. Hence, we can solve for μ_{N-1}. Furthermore, note that μ_N is equal to zero as y_N is a martingale. For a one-factor model (i.e. $\rho_{N,N-1} = 1$) we obtain

$$\mu_{N-1} = -\frac{\alpha_N \sigma_N y_N \alpha_{N-1} \sigma_{N-1} y_{N-1}}{\alpha_{N-1} + \alpha_N(1 + \alpha_{N-1} y_N)}. \tag{8.33}$$

Hence, y_{N-1} follows the process

$$dy_{N-1}(t) = \mu_{N-1}\, dt + \sigma_{N-1}(t) y_{N-1}(t)\, dW^{N+1} \tag{8.34}$$

under the measure $\mathbb{Q}^{N+1}$. Note that μ_{N-1} depends on y_N and is therefore stochastic which implies that y_{N-1} does not have a lognormal distribution under $\mathbb{Q}^{N+1}$.

We can telescope the recursion (8.30) and express the PVBP $P_{n,N+1}$ as

$$\frac{P_{n,N+1}(t)}{D_{N+1}(t)} = \alpha_{n-1} + \sum_{k=n}^{N} \alpha_k \prod_{i=n}^{k} (1 + \alpha_{i-1} y_i(t)). \tag{8.35}$$

From the fact that the left-hand side of (8.35) is a martingale under $\mathbb{Q}^{N+1}$ we can solve for the drift μ_n of $y_n(t)$. However, it will be clear that the expressions for the drift terms are very complicated in the swap market model. We will show how the swap market model can be implemented more elegantly by simulating the martingales $P_{n,N+1}/D_{N+1}$ in Section 8.6.1.

8.2.4 T_1-Forward Measure

In this subsection, we consider an economy with N par swap rates $y_{1,n}$ for $n = 2, \ldots, N + 1$. These swaps have payments on the dates $T_1, \ldots, T_{N+1}$. To simplify the notation, we denote in this subsection $y_{1,n}$ by y_n.

For each swap rate y_n we assume that under the measure $\mathbb{Q}^{2,n}$ its process is given by (8.25). We want to bring all these processes under the T_1-forward measure $\mathbb{Q}^1$.

Like in the previous subsection, we can derive a recursive relation between the PVBP's

$$P_{2,n-1} = P_{2,n} - \alpha_{n-1} D_n$$

$$\Downarrow$$

$$P_{2,n-1} = P_{2,n} + \alpha_{n-1}(D_1 - D_n) - \alpha_{n-1} D_1$$

$$\Downarrow$$

$$P_{2,n-1} = P_{2,n} + \alpha_{n-1} y_n P_{2,n} - \alpha_{n-1} D_1$$

$$\Downarrow$$

$$P_{2,n} = \frac{\alpha_{n-1} D_1 + P_{2,n-1}}{1 + \alpha_{n-1} y_n}. \tag{8.36}$$

Telescoping this recursion we find

$$\frac{P_{2,n}(t)}{D_1(t)} = \sum_{k=2}^{n} \frac{\alpha_{k-1}}{\prod_{i=k}^{n} \left(1 + \alpha_{i-1} y_i(t)\right)}. \tag{8.37}$$

Again, we could solve for the drift terms for all the swap rates from the fact that all $P_{2,n}(t)/D_1(t)$ are martingales under $\mathbb{Q}^1$. However, in Section 8.6.2 we show how swap market models can be implemented more easy via the martingales $P_{2,n}/D_1$.

8.3 Monte Carlo Simulation for LIBOR Market Models

Given the complexity of the processes in the LIBOR and swap market models, prices for other derivatives have to be calculated using numerical methods. One important method that is widely used for market models is *Monte Carlo* simulation.

Table 8.1. Construction of LIBOR rates

	$W^{N+1}(T_1)$	$W^{N+1}(T_2)$	$W^{N+1}(T_3)$	$\cdots$	$W^{N+1}(T_N)$
$L_1(0)$	$L_1(T_1)$				
$L_2(0)$	$L_2(T_1)$	$L_2(T_2)$			
$L_3(0)$	$L_3(T_1)$	$L_3(T_2)$	$L_3(T_3)$		
$\vdots$	$\vdots$	$\vdots$	$\vdots$	$\ddots$	
$L_N(0)$	$L_N(T_1)$	$L_N(T_2)$	$L_N(T_3)$	$\cdots$	$L_N(T_N)$

To implement the LIBOR market model, we work under the terminal measure $\mathbb{Q}^{N+1}$. We proceed along the lines outlined in Chapter 2. First, draw a path for the Brownian motion W^{N+1} at the reset dates T_n using the formula

$$W^{N+1}(T_{n+1}) = W^{N+1}(T_n) + \sqrt{T_{n+1} - T_n}\, \epsilon_n, \tag{8.38}$$

where the ϵ_n are drawings from a standard normal distribution. Based on this path for the Brownian motion we can calculate the forward LIBOR rates given in Table 8.1.

The forward LIBOR rates in the first column are the forward rates observed at time 0. In the subsequent columns we see for each reset date T_n the forward rates that depend on the Brownian motion $W^{N+1}(T_n)$. The forward rates $L_n(T_n)$ are the realisations of the spot LIBOR rates. In each column, the forward rate $L_i(T_{n+1})$ is updated using the discretisation of (8.15)

$$
L_i(T_{n+1}) = L_i(T_n) - \sum_{k=i+1}^{N} \frac{\alpha_k \sigma_k(T_n) L_k(T_n)}{1 + \alpha_k L_k(T_n)} \sigma_i(t) L_i(T_n)(T_{n+1} - T_n)
$$
$$
+ \sigma_i(T_n) L_i(T_n) \big(W^{N+1}(T_{n+1}) - W^{N+1}(T_n) \big). \quad (8.39)
$$

Alternatively, we can use a discretisation based on the stochastic differential equation of $\log(L_i)$ which yields

$$
L_i(T_{n+1}) = L_i(T_n) \exp \left\{ \left(- \sum_{k=i+1}^{N} \frac{\alpha_k \sigma_k(T_n) L_k(T_n)}{1 + \alpha_k L_k(T_n)} \sigma_i(T_n) - \tfrac{1}{2}\sigma_i(T_n)^2 \right) \right.
$$
$$
\left. \times (T_{n+1} - T_n) + \sigma_i(T_n) \big(W^{N+1}(T_{n+1}) - W^{N+1}(T_n) \big) \right\}. \quad (8.40)
$$

Given a set of LIBOR rates realised on this path of the Brownian motion, we can calculate for any time point T_n the value of the discount bonds $D_i(T_n)$ for $i = n, \ldots, N+1$ as

$$
D_i(T_n) = \prod_{k=n}^{i-1} \big(1 + \alpha_k L_k(T_n) \big)^{-1} \quad (8.41)
$$

which is immediate from (8.1).

8.3.1 Calculating the Numeraire Rebased Payoff

Now that we have calculated the LIBOR rates and the discount factors for a particular path, we are in a position to determine the payoff of a given derivative. For example, for a caplet the payoff at time T_{n+1} would be $V(T_{n+1}) = \max\{L_n(T_n) - K, 0\}$.

As we are working under the measure $\mathbb{Q}^{N+1}$, which has D_{N+1} as the numeraire, we have to divide this payoff by the numeraire. This can be done in more than one way, depending on the time at which we consider the payoff. Let $V'(t)$ denote the value at time t of the payoff V divided by the numeraire.

The caplet payoff $V(T_{n+1})$ is determined at time T_n but is not received until time T_{n+1}. Hence, we can calculate the numeraire rebased payoff at time T_{n+1} as

Table 8.2. Path of LIBOR rates

$\Delta T = 0.5, \sigma = 0.15, \alpha = 0.5$

T	$T_0 = 0$	$T_1 = 0.5$	$T_2 = 1.0$	$T_3 = 1.5$	$T_4 = 2.0$
$\Delta W^{(5)}$		-1.15021	1.39659	-0.51945	0.30945
$L_0(T)$	5.000 %				
$L_1(T)$	5.000 %	4.181 %			
$L_2(T)$	5.000 %	4.182 %	5.125 %		
$L_3(T)$	5.000 %	4.183 %	5.128 %	4.715 %	
$L_4(T)$	5.000 %	4.184 %	5.130 %	4.719 %	4.916 %
$D_1(T)$	0.97561				
$D_2(T)$	0.95181	0.97952			
$D_3(T)$	0.92860	0.95946	0.97502		
$D_4(T)$	0.90595	0.93981	0.95064	0.97697	
$D_5(T)$	0.88385	0.92055	0.92687	0.95445	0.97601

$$V'(T_{n+1}) = \frac{V(T_{n+1})}{D_{N+1}(T_{n+1})} \tag{8.42}$$

and at time T_n as

$$V'(T_n) = \frac{V(T_{n+1})D_{n+1}(T_n)}{D_{N+1}(T_n)}. \tag{8.43}$$

Note that for a given path of LIBOR rates these two expressions need not be the same, but in the LIBOR market model these expressions are equivalent in the sense that $V'(T_n) = \mathbb{E}^{N+1}\left(V'(T_{n+1}) \mid \mathcal{F}_{T_n}\right)$.

We can also reinvest the payoff from time T_{n+1} until time T_{N+1} into a deposit, thereby earning the LIBOR interest for each period. This leads to the expression

$$V(T_{N+1}) = V(T_{n+1}) \prod_{k=n+1}^{N} \left(1 + \alpha_k L_k(T_k)\right). \tag{8.44}$$

Again, this expression is equivalent to the two previous expressions in the sense that $V'(T_n) = \mathbb{E}^{N+1}\left(V'(T_{N+1}) \mid \mathcal{F}_{T_n}\right)$.

8.3.2 Example: Vanilla Cap

In this section, we show how to construct a single path of LIBOR rates. Then we calculate the numeraire rebased payoff of a vanilla cap for this particular path.

We take a LIBOR market model with semi-annual LIBOR rates. For simplicity we assume $T_i = 0.5i$ for $i = 0, \ldots, N = 4$ and $\alpha_i \equiv 0.5$. Also we assume a flat initial term-structure of LIBOR rates at 5% and a flat term-structure of volatility at 15%.

First we have to draw a path of increments $\Delta W^{(5)}$ of the Brownian motion under the measure $\mathbb{Q}^5$ using (8.38), then we can calculate the forward LIBOR rates using (8.40). Finally, we compute the discount factors using (8.41). The results for one particular path are reported in Table 8.2.

Because we start with a flat yield-curve and because we work in a one-factor model, we see that the different forward LIBOR rates move almost in parallel. We do see the impact of the drift terms under the terminal measure as the forward LIBOR rates are slightly different along a column. Along the diagonal we can read the fixings of the spot LIBOR rates.

The value V^{cap} of a vanilla cap with a strike of 5% is the sum of the numeraire rebased caplet payoffs along this path. For this particular path only $L_2(T_2) = 5.125\%$ is in-the-money. The numeraire rebased value of this caplet can, for this path, be calculated as

$$V'(T_2) = (L_2(T_2) - 5\%)D_3(T_2)/D_5(T_2) = 0.001315 \tag{8.45}$$

or

$$V'(T_3) = (L_2(T_2) - 5\%)/D_5(T_3) = 0.001310 \tag{8.46}$$

or

$$V'(T_5) = (L_2(T_2) - 5\%)(1 + 0.5L_3(T_3))(1 + 0.5L_4(T_4))$$
$$= 0.001311. \tag{8.47}$$

We know from Section 8.3.1 that all these methods are equivalent.

If we would draw 100,000 paths and calculate for each path j the numeraire rebased value $V^{\text{cap}}{}'_j$, then we would estimate the Monte Carlo simulation value of the cap as $V^{\text{cap}} = D_5(0)(\sum_{j=1}^{100,000} V^{\text{cap}}{}'_j)/100,000$.

8.3.3 Discrete Barrier Caps/Floors

Discrete barrier options are exotic options with a payoff conditional on the event that past LIBOR rates have not reached a certain level known as the *barrier*. Examples of these options, in the case of interest rates, are discrete barrier caps, floors and digitals, where the barrier condition is reviewed every fixing date. First we review different types of instruments that are actually traded in the market. Then we show how these instruments can be valued using Monte Carlo simulation in a LIBOR market model.

The discrete barrier options we are considering have the barrier condition monitored only at discrete points in time. For the case of interest rate options we are considering here, the barrier is only checked when a new LIBOR rate is set. Just as with regular caps and floors, we can decompose a discrete barrier cap/floor as a portfolio of discrete barrier caplets/floorlets. Hence, the product is closely related to a standard cap. In its fully general form, we have associated with each caplet i the strike K_i and a barrier level H_i. However, to obtain a closer link with the regular traded caps we use in our examples a single strike K and barrier H for the whole deal.

To value a down and out discrete barrier cap we can use Monte Carlo simulation in the LIBOR market model as described in Section 8.3. Under the terminal measure $\mathbb{Q}^{N+1}$ we use D_{N+1} as numeraire. Given that we have drawn a path with the spot LIBOR rates $L_n(T_n)$ the value of the ith up and out discrete barrier caplet depends on all the LIBOR rates L_n with $n \leq i$. The numeraire rebased value V' for this path is given by

$$V' = \frac{\max\{L_i(T_i) - K, 0\} \prod_{n=1}^{i} \mathbb{1}(L_n(T_n) > H)}{D_{N+1}(T_{i+1})}, \qquad (8.48)$$

where $\mathbb{1}()$ denotes the indicator function which is 1 if the argument is true and 0 if the argument is false. This formula denotes that the caplet payoff $\max\{L_i - K, 0\}$ is received only if all previous LIBOR rates did fix above the barrier H.

Let us consider for example a down and out caplet with barrier 4.25% and strike 4.75%. If we use the path of LIBOR rates of Table 8.2 then we see that an ordinary caplet that is fixed at time T_4 has a payoff of $(0.04916 - 0.04750)$ at time T_5. However, the down and out barrier caplet has a payoff of 0, because L_1 was fixed below the barrier at time T_1.

The payoff of a down and in discrete barrier cap is given by

$$V' = \frac{\max\{L_i(T_i) - K, 0\} \left(1 - \prod_{n=1}^{i} \mathbb{1}(L_n(T_n) > H)\right)}{D_{N+1}(T_{i+1})}. \qquad (8.49)$$

This formula denotes that the caplet payoff is received if at least one of the previous LIBOR rates set above the barrier. Note that the combination of a up and in plus an up and out barrier caplet is equal to an ordinary caplet.

By combining the following features: in/out, up/down and cap/floor we can create eight different variants of discrete barrier caps and floors.

In Table 8.3 below we denote the price in basispoints obtained by simulation in the LIBOR market model by "LMM" and the standard error of the simulation by "SE". For the simulations we used a yield-curve given by the zero curve $Z(T) = 0.08 - 0.05 \exp\{-0.18T\}$. The prices of the discount bonds at time 0 are thus given by $D_n(0) = \exp\{-Z(T_n)T_n\}$. To calculate the LIBOR fixing dates T_n we made the simplifying assumption that $T_n = 0.5n$ for semi-annual payments, with daycount fractions α_n being constant and equal to 0.5. The time-steps used in the Monte Carlo simulation are also equal to 0.5.

From the table we see first of all that even with the very basic Monte Carlo procedure described in this book, we already get prices accurate to within one basispoint for simulations with 100,000 paths. This accuracy can be improved even further by using variance reduction techniques. Also we see that the up and out cap and floor prices converge to the normal cap and floor prices if the barrier becomes higher as it is less and less likely that the barrier is hit by one of the LIBOR rates.

Table 8.3. Prices of discrete barrier caps/floors

Semi-Annual, $M = 100,000$, Flat Vol 10.00%

Mat	Strike	Barrier	LMM	(SE)
Up & Out Cap				
2Y	4.00%	5.00%	37.69	(0.08)
2Y	4.00%	7.00%	196.80	(0.22)
3Y	4.50%	5.50%	39.60	(0.09)
3Y	4.50%	7.50%	247.40	(0.31)
5Y	5.00%	6.00%	44.04	(0.11)
5Y	5.00%	8.00%	334.11	(0.48)
7Y	5.50%	6.50%	45.92	(0.12)
7Y	5.50%	8.50%	366.84	(0.61)
10Y	6.00%	7.00%	46.79	(0.13)
10Y	6.00%	9.00%	387.00	(0.74)
Up & Out Floor				
2Y	4.00%	3.00%	0.00	(0.00)
2Y	4.00%	5.00%	1.57	(0.01)
2Y	4.00%	7.00%	1.57	(0.01)
3Y	4.50%	3.50%	0.16	(0.01)
3Y	4.50%	5.50%	16.06	(0.06)
3Y	4.50%	7.50%	16.06	(0.06)
5Y	5.00%	4.00%	11.77	(0.08)
5Y	5.00%	6.00%	52.47	(0.13)
5Y	5.00%	8.00%	52.51	(0.13)
7Y	5.50%	4.50%	65.32	(0.18)
7Y	5.50%	6.50%	114.80	(0.27)
7Y	5.50%	8.50%	115.25	(0.27)
10Y	6.00%	5.00%	143.48	(0.35)
10Y	6.00%	7.00%	215.86	(0.55)
10Y	6.00%	9.00%	219.38	(0.55)

8.3.4 Discrete Barrier Digital Caps/Floors

A discrete barrier digital cap/floor is an instrument that has the same characteristics as a normal discrete barrier cap/floor, except that the payoff is a fixed amount paid if the final LIBOR rate is above the strike (digital cap) or below the strike (digital floor). Again, a discrete barrier digital cap/floor can be decomposed as a portfolio of barrier digital optionlets. The numeraire rebased payoff of an up and out digital caplet can be represented as

$$V' = \frac{\mathbb{1}(L_i(T_i) - K > 0) \prod_{n=1}^{i} \mathbb{1}(L_n(T_n) < H)}{D_{N+1}(T_{i+1})}. \qquad (8.50)$$

8.3.5 Payment Stream

A payment stream is a contract that pays a constant amount for every LIBOR reset for which the barrier was not yet hit. This product can also be modelled as an out barrier digital cap with a strike of zero, or as an out barrier digital floor with a strike at $+\infty$.

8.3.6 Ratchets

A *ratchet option* comprises a number of payment dates $T_2, \ldots, T_{N+1}$. Each coupon payment is similar to a vanilla cap except that the strike is variable and depends on earlier LIBOR resets. The payment at time T_{i+1} is for an amount

$$V_i = A_i \alpha_i \max\{\phi_i(L_i(T_i) - K_i), 0\}, \tag{8.51}$$

where K_1 is given and for $i > 1$

$$K_i = K(V_{i-1}, \ldots, V_1; L_{i-1}, \ldots, L_1). \tag{8.52}$$

In the above, the constants A_i are fixed notionals, $L_i(T_i)$ are LIBOR settings, α_i are daycount fractions and ϕ_i are $+1/-1$ variables indicating call or put.

The example of the "classical" ratchet option (also known as a *sticky cap*) is a bond where the coupons are based on LIBOR rates, but if the LIBOR setting is lower than the previous coupon, the previous coupon is retained. Hence, the coupon payments can only go up, which explains the name ratchet. The payments can be expressed as

$$V_i = \max\{L_i, V_{i-1}\} = L_i + \max\{V_{i-1} - L_i, 0\}. \tag{8.53}$$

Using the path of LIBOR rates from Table 8.2 we would receive the following payments $5.000\%, 5.000\%, 5.125\%, 5.125\%, 5.125\%$ at times $T_1, \ldots, T_5$ respectively.

Another example is a bond that pays a coupon every 6 months. Each coupon is 6M LIBOR + 35bp but with a maximum value equal to the previous coupon plus 40bp. Thus the owner of one of these bonds has effectively sold a ratchet option with each option payment equal to

$$V_i = \max\{L_i + 0.0035 - (V_{i-1} + 0.0040), 0\} \tag{8.54}$$

and $V_1 = L_1 + 0.0035$. Using the LIBOR rates from Table 8.2 the bond pays the coupons $5.350\%, 4.531\%, 4.931\%, 5.065\%, 5.266\%$ at times $T_1, \ldots, T_5$ respectively.

To calculate the value of a ratchet option, we use Monte Carlo simulation of the LIBOR market model. For each path of LIBOR rate settings we generate, we can calculate the coupon payments. Each coupon payment needs to be divided with the value of the numeraire discount bond for this particular path at the time the coupon is received. Then the option value is determined by taking the average over all paths.

8.4 Monte Carlo Simulation for Swap Market Models

To value swap based derivatives other than European-style swaptions, we can use Monte Carlo simulation in the swap market model. To implement we could simulate the forward swap rates directly under a single measure. However, as we saw in Section 8.2, the expressions for the drift terms are quite complicated in the swap market model. We discuss therefore in this section an alternative approach to simulate swap market models which is easier to implement.

8.4.1 Terminal Measure

Under the terminal measure we can exploit the fact that the numeraire re-based PVBP $P'_{n,N+1} = P_{n,N+1}/D_{N+1}$ is a martingale under the measure $\mathbb{Q}^{N+1}$. We can apply Itô's Lemma to obtain

$$dP'_{n,N+1} = \sum_{l=n}^{N} \left(\frac{\partial P'_{n,N+1}}{\partial y_l} \right) \sigma_l y_l \, dW^{N+1}, \qquad (8.55)$$

where y_l denotes $y_{l,N+1}$. Substituting equation (8.35) and evaluating the partial derivatives yields

$$dP'_{n,N+1} = \sum_{l=n}^{N} \left(\sum_{k=l}^{N} \alpha_k \frac{\alpha_{l-1}}{1+\alpha_{l-1}y_l} \prod_{i=n}^{k} (1+\alpha_{i-1}y_i) \right) \sigma_l y_l \, dW^{N+1}. \qquad (8.56)$$

We can split the outer summation in a term for $l = n$ and the remaining terms for $l = n+1, \ldots, N$. Simplifying then gives

$$dP'_{n,N+1} = \alpha_{n-1} P'_{n+1,N+1} \sigma_n y_n dW^{N+1} + (1+\alpha_{n-1}y_n)dP'_{n+1,N+1}, \qquad (8.57)$$

where $P'_{N+1,N+1} \equiv \alpha_N$. Equation (8.57) is easy to discretise and use in a Monte Carlo simulation.

We illustrate how to draw a single path with an example. We take a swap market model with semi-annual payment dates. For simplicity we assume $T_i = 0.5i$ for $i = 0, \ldots, N = 4$ and $\alpha_i \equiv 0.5$. Also we assume a flat initial term-structure of swap rates at 5% and a flat term-structure of volatility at 15%. Hence, we model the behaviour of the forward swap rates $y_1, \ldots, y_4$, the PVBP's $P_{1,5}, P_{2,5}, \ldots, P_{5,5}$ and we use D_5 as the numeraire.

With the initial forward swap rates, we can calculate the initial PVBP's. Then we can update the numeraire rebased PVBP values for the next time step using (8.57) inductively for $n = N, \ldots, 1$. From the updated PVBP's we can calculate the (numeraire rebased) discount factors and the forward swap

Table 8.4. Path of swap rates under terminal measure

$$\Delta T = 0.5, \sigma = 0.15, \alpha = 0.5$$

T	$T_0 = 0$	$T_1 = 0.5$	$T_2 = 1.0$	$T_3 = 1.5$	$T_4 = 2.0$
$\Delta W^{(5)}$		-1.15021	1.39659	-0.51945	0.30945
$y_{1,5}$	5.000%	4.132%			
$y_{2,5}$	5.000%	4.134%	4.996%		
$y_{3,5}$	5.000%	4.136%	5.000%	4.610%	
$y_{4,5}$	5.000%	4.137%	5.004%	4.614%	4.828%
$P_{1,5}/D_5$	2.62816	2.60550			
$P_{2,5}/D_5$	2.07626	2.06288	2.07624		
$P_{3,5}/D_5$	1.53781	1.53123	1.53782	1.53485	
$P_{4,5}/D_5$	1.01250	1.01034	1.01251	1.01154	1.01207
$P_{5,5}/D_5$	0.50000	0.50000	0.50000	0.50000	0.50000
D_1/D_5	1.10381	1.08523			
D_2/D_5	1.07689	1.06330	1.07683		
D_3/D_5	1.05063	1.04178	1.05063	1.04663	
D_4/D_5	1.02500	1.02069	1.02502	1.02307	1.02414

rates. Then we can proceed to the next time-point. An example path is shown in Table 8.4.

A typical class of swap derivatives one would like to model are *Bermudan swaptions*. A Bermudan swaption holder has the right at different points in time to enter into an underlying swap. However, as is well known American- and Bermudan-style products are extremely difficult to value accurately with Monte Carlo simulation. Some progress has been made recently (see Pedersen (1999) for a comparison of different methods to value Bermudan swaptions in a market model framework with Monte Carlo simulation) but the methods proposed are still very complicated. In Chapter 9 we present a class of models which are particularly suited for pricing Bermudan-style interest rate derivatives efficiently.

8.4.2 T_1-Forward Measure

To value swap based derivatives under the T_1-forward measure, we exploit the fact that the numeraire rebased PVBP $P'_{2,n} = P_{2,n}/D_1$ is a martingale under the measure $\mathbb{Q}^1$. In this case we make the assumption that the forward swap rates $y_l = y_{1,l}$ follow the processes $dy_l = \sigma_l y_l dW_l^{2,l}$ under the measure $\mathbb{Q}^{2,l}$ for $l = 2, \ldots, N+1$. Note that we have introduced a multi-factor model here where each forward swap rate y_l is driven by its own Brownian motion W_l. We assume that the instantaneous correlations between the Brownian motions $dW_j \, dW_k = \rho_{jk} \, dt$ are given. We can apply Itô's Lemma to obtain

$$dP'_{2,n} = \sum_{l=2}^{n} \left(\frac{\partial P'_{2,n}}{\partial y_l} \right) \sigma_l y_l \, dW_l^{(1)}. \tag{8.58}$$

Evaluating the partial derivatives using equation (8.37) yields

$$dP'_{2,n} = \sum_{l=2}^{n} \left(\sum_{k=2}^{l} \alpha_{k-1} \frac{-\alpha_{l-1}}{1+\alpha_{l-1}y_l} \prod_{i=k}^{n} (1+\alpha_{i-1}y_i)^{-1} \right) \sigma_l y_l \, dW_l^{(1)}, \quad (8.59)$$

which can be simplified to

$$dP'_{2,n} = \frac{dP'_{2,n-1}}{(1+\alpha_{n-1}y_n)} - \frac{\alpha_{n-1}}{(1+\alpha_{n-1}y_n)} P'_{2,n} \sigma_n y_n \, dW_n^{(1)}. \quad (8.60)$$

As a boundary condition we can use $P'_{2,1} \equiv 0$.

Given the initial forward swap rates, we can calculate the initial PVBP's. Then we can update the numeraire rebased PVBP values for the next time step using (8.60) inductively for $n = 2, \ldots, N+1$. From the updated PVBP's we can calculate the (numeraire rebased) discount factors and the forward swap rates. We can then proceed to the next time-point.

8.4.3 Example: Spread Option

A typical class of products where one would use a swap market model under the T_1-forward measure are *spread options*. In a spread option two rates or value are observed and the difference between those rates is paid out. Suppose we have two forward swap rates y_a and y_b.

In a *rate based* spread option both rates are observed at time T_1 and a payoff of $\max\{y_a(T_1) - y_b(T_1) - K, 0\}$ is made at time $S \geq T_1$. Hence, this option gives the right to choose the better of two interest rates.

In a *value based* spread option two swap values are observed at time T_1 and a payoff of $\max\{P_a(T_1)(y_a(T_1) - K_a) - P_b(T_1)(y_b(T_1) - K_b) - K, 0\}$ is paid at time $S \geq T_1$. This option gives the right to enter into the better of two swap contracts.

Because spread options depend on multiple interest rates observed at the same time-point, a multi-factor interest rate model must be used to obtain an accurate price. Hence, we will construct an example of a two-factor swap market model. We consider six semi-annual time-steps $S_i = 0.5i$ and also set $\alpha \equiv 0.5$. We only model two swap rates $y_5 = y_{S_4, S_5}$ and $y_6 = y_{S_4, S_6}$ with $y_5(0) = y_6(0) = 5\%$. We will use D_{S_4} as the numeraire, hence $T_1 = S_4$. Finally we assume that the instantaneous correlation between the driving Brownian motions is $dW_5 \, dW_6 = 0.5 \, dt$.

First question we have to address is how to generate Brownian motions that are correlated. The case of two Brownian motions is straightforward. Suppose we start with two independent Brownian motions W_a and W_b. Then we can construct W_5 and W_6 with correlation ρ as follows

$$W_5 = W_a,$$
$$W_6 = \rho W_a + \sqrt{1 - \rho^2} W_b. \quad (8.61)$$

Table 8.5. Path of swap rates under T_1-forward measure

$\Delta T = 0.5, \sigma = 0.15, \alpha = 0.5, \rho = 0.5, T_1 = S_4$

	$S_0 = 0$	$S_1 = 0.5$	$S_2 = 1.0$	$S_3 = 1.5$	$S_4 = 2.0$
ΔW_a		-0.42930	0.74289	0.22289	1.02136
ΔW_b		-0.21569	-0.42968	0.11661	-0.17149
ΔW_5		-0.42930	0.74289	0.22289	1.02136
ΔW_6		-0.40145	-0.00067	0.21243	0.36216
y_5	5.000%	4.679%	5.201%	5.375%	6.202%
y_6	5.000%	4.700%	4.699%	4.849%	5.113%
P_{54}/D_4	0.00000	0.00000	0.00000	0.00000	0.00000
P_{55}/D_4	0.48780	0.48857	0.48733	0.48691	0.48496
P_{56}/D_4	0.96371	0.96587	0.96466	0.96355	0.96041
D_5/D_4	0.97561	0.97714	0.97465	0.97383	0.96992
D_6/D_4	0.95181	0.95461	0.95467	0.95328	0.95089

This procedure is known a Cholesky decomposition of the correlation matrix and can be generalised to higher dimensions (see Press et al. (1992, Chapter 2.9)).

For our example, we have generated a single path in the swap market model. The results are reported in Table 8.5. First thing to note is that the simulated path only runs until S_4. At this time point the numeraire discount bond D_4 reaches maturity and ceases to exist after this time point. However, because the payoff of the spread option products is determined at this time, we have sufficient information to determine the payoff of different spread options. For example: (All variables to be read from column S_4, also note that $D_4(S_4) \equiv 1$.)

- rate based spread option paid at S_4: $\max\{y_5 - y_6 - K, 0\}$;
- rate based spread option paid at S_5: $\max\{y_5 - y_6 - K, 0\}D_5$;
- rate based spread option paid as fixed rate on S_5 and S_6:
  $\max\{y_5 - y_6 - K, 0\}P_{56}$;
- value based spread option paid at S_4:
  $\max\{P_{55}(y_5 - K_5) - P_{56}(y_6 - K_6), 0\}$;
- value based spread option paid at S_6:
  $\max\{P_{55}(y_5 - K_5) - P_{56}(y_6 - K_6), 0\}D_6$.

For an alternative method to value spread option products we refer to Chapter 11, Section 11.4.

9. Markov-Functional Models

In this chapter we propose models that can fit the observed prices of liquid instruments in a similar fashion to the market models, but which also have the advantage that prices can be calculated just as efficiently as in the spot rate models of the first part of this book. To achieve this we consider the general class of *Markov-Functional* interest rate models (MF models), first introduced by Hunt, Kennedy and Pelsser (2000). The defining characteristic of MF models is that pure discount bond prices are assumed at any time to be a function of some low-dimensional process which is Markovian in some martingale measure. This ensures that implementation is efficient since it is only necessary to track the driving Markov process. Market models do not possess this property (for a *low*-dimensional Markov process) and this is the impediment to their efficient implementation. The freedom to choose the functional form is what permits accurate calibration of Markov-Functional models to relevant market prices, a property not possessed by spot rate models. The remaining freedom to specify the law of the driving Markov process is what allows us to make the model realistic. As we shall see, given a Markov process, it is possible to fit the *marginal distributions* of market interest rates implied by the market option prices. However, for successfully pricing exotic interest rate products it is important to capture the *joint distribution* of the interest rates under consideration. We will discuss this important point further in Section 9.5.

The rest of this chapter is organised as follows. In Section 9.1 we discuss the basic assumptions of the MF model. The construction of the LIBOR and swap MF model and details on the numerical implementation of the models are discussed in Sections 9.2, 9.3 and 9.4. In Section 9.5 the impact of mean-reversion on the auto-correlation of the interest rates is discussed. Finally, we illustrate the use of MF models with the examples of barrier caps, auto- and chooser-caps and Bermudan swaptions in Sections 9.6, 9.7 and 9.8 respectively.

9.1 Basic Assumptions

The assumptions we are going to make here are motivated by two key issues: first, the need for a model to be well-calibrated to market prices of relevant standard market instruments and, secondly, the requirement that the model can be efficiently implemented.

Central to the approach is the assumption that the state of the economy at t is summarised via some low dimensional (time-inhomogeneous) Markov process $x(t)$. This is clearly the defining property of a model that can be implemented in practice. It is true for all classical spot rate models, in which case $x(t) = r(t)$. Models not possessing this property are market models of the previous chapter, where the dimension of the Markov process x is much higher.

With the exception of market models, which suffer from high dimensionality, existing models fail to calibrate well to the distribution of relevant market rates. The key to achieving this, without the extra dimensionality, is first to define a process x of low dimension and then to define its relationship to the assets in the economy in a way which yields the desired distributions. In practice x will be one or at most two dimensional, the examples in this chapter all being one dimensional.

Like in the market models, we choose a tenor structure $T_1, \ldots, T_{N+1}$ and we take the terminal discount bond $D_{N+1} = D_{T_{N+1}}$ as numeraire. Hence we will work under the martingale measure $\mathbb{Q}^{N+1}$.

Let us now make an assumption about the underlying Markov process x. A convenient choice is

$$dx(t) = \tau(t)\, dW^{N+1}, \tag{9.1}$$

where $\tau(t)$ is a deterministic function and W^{N+1} denotes Brownian motion under $\mathbb{Q}^{N+1}$. For this choice we know the law of x analytically. Conditional on $x(t)$ the random variable $x(s)$ has for $s \geq t$ a normal probability distribution with mean $x(t)$ and variance $\int_t^s \tau^2(u)\, du$. The probability density function of $x(s)$ given $x(t)$ is denoted by $\phi\big(x(s)|x(t)\big)$ and can be expressed as

$$\phi\big(x(s)|x(t)\big) = \frac{\exp\left(-\frac{1}{2}\frac{(x(t)-x(s))^2}{\int_t^s \tau^2(u)\, du}\right)}{\sqrt{2\pi \int_t^s \tau^2(u)\, du}}. \tag{9.2}$$

We will explain the impact of different choices for $\tau(t)$ in Section 9.5.

The crucial assumption on which the MF model is built is the assumption that the numeraire discount bond is a function $D_{N+1}\big(t, x(t)\big)$ monotonic in x. This means that we assume that the value of the numeraire discount bond at time t is determined completely by the process $x(t)$.

This is, of course, very similar to the situation we have for spot rate models where the value of the discount bond is determined by the value of the spot interest rate $r(t)$. The big difference with spot rate models is that the functional forms of the discount bonds will not be determined endogenously by the model (like in the case of Hull-White models or Squared Gaussian models) but will be determined by the market prices of caplets or swaptions.

Note, that once we have obtained the functional form of the numeraire discount bond, we can determine the value of all other discount bonds D_S with $S \leq T_{N+1}$ using

$$\frac{D_S(t)}{D_{N+1}(t)} = \mathbb{E}^{N+1}\left(\frac{D_S(S)}{D_{N+1}(S)}\,\middle|\,\mathcal{F}_t\right) = \mathbb{E}^{N+1}\left(\frac{1}{D_{N+1}(S)}\,\middle|\,\mathcal{F}_t\right) \quad (9.3)$$

$$= \int_{-\infty}^{\infty} \frac{1}{D_{N+1}(S,z)}\phi\big(z|x(t)\big)\,dz. \quad (9.4)$$

Equation (9.4) shows how the conditional expectation (9.3) can be calculated as an integral using the density function ϕ of the process x where the integration dummy z goes through all the possible values of $x(S)$. Because we have assumed that all information in the economy is reflected by the Markov process x, conditioning on $x(t)$ is equivalent to conditioning on $\mathcal{F}_t$. Once the discount bonds are known, we can calculate the price of any payoff we are interested in.

The only thing that remains to be done, is to determine the functional form of the numeraire discount bond by fitting it to the prices of market instruments. We can either fit to caplets, which leads to a LIBOR MF model or fit to swaptions which leads to a swap MF model.

The basic idea of the fitting procedure is as follows. Suppose we are given at time 0 market prices for all possible strikes K for a reference instrument $V_i^{\text{ref}}(K)$ (like a caplet or a swaption) that matures at T_i. At maturity the payoff of the instrument is a function $V_i^{\text{ref}}\big(D_{N+1}(T_i), K\big)$ of the discount bond D_{N+1}. Under the measure $\mathbb{Q}^{N+1}$ we have

$$\frac{V_i^{\text{ref}}(K)}{D_{N+1}(0)} = \mathbb{E}^{N+1}\left(\frac{V_i^{\text{ref}}\big(D_{N+1}(T_i), K\big)}{D_{N+1}(T_i)}\right)$$
$$= \int_{-\infty}^{\infty} \frac{V_i^{\text{ref}}\big(D_{N+1}(T_i, z), K\big)}{D_{N+1}(T_i, z)}\phi\big(z|x(0)\big)\,dz. \quad (9.5)$$

This is a (non-linear) integral equation in D_{N+1} which can be solved to obtain $D_{N+1}(T_i)$. In the sections below we explain how to use this basic idea to fit D_{N+1} to the prices of caplets or swaptions.

9.2 LIBOR Markov-Functional Model

Like in the LIBOR market model, we assume a set of N LIBOR rates $L_i = L_{T_i, T_{i+1}}$ that matches the tenor structure $\{T_i\}$. We also assume that the market prices for the caplets on these LIBOR rates are given by the Black (1976) formula with implied volatilities $\bar{\sigma}_i$. Note that this assumption can be re-laxed to allow for any probability distribution implied by the market, but for expositional purposes we make the lognormal assumption here.

To solve for the functional forms of $D_{N+1}(T_i)$ we work backwards from time T_{N+1}. At time T_{N+1} we know that $D_{N+1}(T_{N+1}) \equiv 1$.

At time T_N we know that $D_{N+1}(T_N) = 1/(1+\alpha_N L_N(T_N))$. We know that given the assumption of the Black formula for the caplet price, the LIBOR rate L_N is a lognormal martingale with volatility $\bar{\sigma}_N$ under the measure $\mathbb{Q}^{N+1}$. Given (9.1) we can express the LIBOR rate explicitly in terms of the Markov process x as

$$L_N(T_N) = L_N(0) \exp\left(-\tfrac{1}{2}\bar{\sigma}_N^2 T_N + \sqrt{\frac{\bar{\sigma}_N^2 T_N}{\int_0^{T_N} \tau^2(s)\, ds}} x(T_N)\right). \qquad (9.6)$$

Note that we set $x(0) = 0$. Hence, we obtain the following explicit expression for D_{N+1}

$$D_{N+1}\big(T_N, x(T_N)\big) = \cfrac{1}{1 + \alpha_N L_N(0) \exp\left(-\tfrac{1}{2}\bar{\sigma}_N^2 T_N + \sqrt{\dfrac{\bar{\sigma}_N^2 T_N}{\int_0^{T_N} \tau^2(s)\, ds}} x(T_N)\right)}. \qquad (9.7)$$

Note that D_{N+1} is a monotonic decreasing function of x.

Let us now assume we have already determined the functional forms of D_{N+1} for $T_N, T_{N-1}, T_{N-2} \ldots, T_{n+1}$. We now show how to find the functional form of D_{N+1} at time T_n. We will proceed in two steps. First we determine the functional form of the LIBOR rate. Then we determine the functional form of D_{N+1} from the LIBOR rate like we did for time T_N.

To find the functional form of the LIBOR rate at time T_n, we use the current market price of the caplets $V_n^{\text{capl}}(K)$. From Dupire (1994) we know that the price of a digital caplet is given by $V_n^{\text{dcl}}(K) = -(\partial/\partial K)V_n^{\text{capl}}(K)$. If we assume the market price of the caplets is given by the Black formula with volatility $\bar{\sigma}_n$, we obtain for the digital caplet price the expression

$$V_n^{\text{dcl}}(K) = D_{n+1}(0)\, N\left(\frac{\log(L_n(0)/K) - \tfrac{1}{2}\bar{\sigma}_n^2 T_n}{\bar{\sigma}_n \sqrt{T_n}}\right), \qquad (9.8)$$

where $N()$ denotes the cumulative normal distribution function.

The value of the digital caplet at time T_n can be expressed as

$$V_n^{\text{dcl}}(T_n, K) = D_{n+1}(T_n)\mathbb{1}(L_n(T_n, x(T_n)) > K). \qquad (9.9)$$

If we assume that the LIBOR rate is a monotonic increasing function of x, then there is a unique value x^* such that $L_n(T_n, x^*) = K$. Hence, we can rewrite the digital caplet value as

$$V_n^{\text{dcl}}(T_n, K) = D_{n+1}(T_n)\mathbb{1}(x(T_n) > x^*). \qquad (9.10)$$

Under the terminal measure $\mathbb{Q}^{N+1}$ the digital caplet value can be expressed as

$$V_n^{\text{dcl}}(K) = D_{N+1}(0)\mathbb{E}^{N+1}\left(\mathbb{1}(x(T_n) > x^*)\frac{D_{n+1}(T_n, x(T_n))}{D_{N+1}(T_n, x(T_n))}\right)$$

$$= D_{N+1}(0)\int_{x^*}^{\infty}\frac{D_{n+1}(T_n, x_n)}{D_{N+1}(T_n, x_n)}\phi(x_n|x(0))\,dx_n$$

$$= D_{N+1}(0)\int_{x^*}^{\infty}\left(\int_{-\infty}^{\infty}\frac{1}{D_{N+1}(T_{n+1}, x_{n+1})}\phi(x_{n+1}|x_n)\,dx_{n+1}\right)$$
$$\times\,\phi(x_n|x(0))\,dx_n. \tag{9.11}$$

The functional form $D_{N+1}(T_{n+1})$ in the last line has already been determined in the previous iteration. Hence, at time T_n we can evaluate (at least numerically) (9.11) for different values of x^*. Let us denote this (numerical) value by $J_n(x^*)$.

Remember that x^* is the value of $x(T_n)$ such that $L_n(T_n, x^*) = K$. Hence, we find from (9.11) $J_n(x^*) = V_n^{\text{dcl}}(K) = V^{\text{dcl}}(L_n(T_n, x^*))$ which is an equation in $L_n(T_n, x^*)$ that gives us the functional form of the LIBOR rate. In the case of the Black formula (9.8), we can solve this expression explicitly to find

$$L_n(T_n, x^*) = L_n(0)\exp\left(-\tfrac{1}{2}\bar{\sigma}_n^2 T_n - \bar{\sigma}_n\sqrt{T_n}\,N^{-1}\left(\frac{J_n(x^*)}{D_{n+1}(0)}\right)\right), \tag{9.12}$$

where $N^{-1}()$ denotes the inverse cumulative normal distribution function.

With the functional form of the LIBOR rate, we can determine the functional form of the numeraire $D_{N+1}(T_n)$ as follows. The LIBOR rate can be written as

$$L_n(T_n) = \frac{1 - D_{n+1}(T_n)}{\alpha_n D_{n+1}(T_n)} = \frac{\frac{1}{D_{N+1}(T_n)} - \frac{D_{n+1}(T_n)}{D_{N+1}(T_n)}}{\alpha_n\frac{D_{n+1}(T_n)}{D_{N+1}(T_n)}}. \tag{9.13}$$

Solving for D_{N+1} gives

$$D_{N+1}(T_n, x^*) = \left((1 + \alpha_n L_n(T_n, x^*))\frac{D_{n+1}(T_n, x^*)}{D_{N+1}(T_n, x^*)}\right)^{-1}, \tag{9.14}$$

where the term D_{n+1}/D_{N+1} has already been calculated in (9.11).

With the procedure outlined above, we can solve the functional forms of the numeraire discount bonds iteratively backwards from time T_{N+1} to time T_1.

9.3 Swap Markov-Functional Model

Another possibility to determine the functional forms of the discount bonds is to fit to the prices of (digital) swaptions. We treat the case of the set of options on swaps with a fixed end-date.

Let T_{N+1} denote the terminal date and let y_n denote the par swap rate of a swap that starts at time T_n and has fixed payments precisely at times $T_{n+1}, T_{n+2}, \ldots, T_{N+1}$. We choose the discount bond D_{N+1} as the numeraire and work under the associated measure $\mathbb{Q}^{N+1}$. The final swap rate y_N is equal to the LIBOR rate L_N. Under the assumption that the market prices are given by the Black formula, the functional form of the numeraire at time T_N is therefore given by (9.7).

Let us now assume we have already determined the functional forms of D_{N+1} for $T_N, T_{N-1}, T_{N-2} \ldots, T_{n+1}$. We now show how to find the functional form of D_{N+1} at time T_n. We will proceed in two steps, first we determine the functional form of the par swap rate then we determine the functional form of D_{N+1} from the swap rate.

To find the functional form of the swap rate at time T_n, we use the current market price of digital swaptions $V_n^{\mathrm{dsw}}(K)$. If we assume the market price of the swaptions is given by the Black formula with volatility $\bar{\sigma}_n$, we obtain for the digital swaption price the expression

$$V_n^{\mathrm{dsw}}(K) = P_n(0) \, N\left(\frac{\log(y_n(0)/K) - \frac{1}{2}\bar{\sigma}_n^2 T_n}{\bar{\sigma}_n \sqrt{T_n}} \right), \qquad (9.15)$$

where P_n denotes the PVBP $P_n(t) = \sum_{k=n+1}^{N+1} \alpha_{k-1} D_k(t)$.

If we assume that the swap rate is a monotonic increasing function of x, then there is a unique value x^* such that $y_n(T_n, x^*) = K$. Under the terminal measure $\mathbb{Q}^{N+1}$ the digital swaption value can be expressed as

$$V_n^{\mathrm{dsw}}(K) = D_{N+1}(0)\mathbb{E}^{N+1}\left(\mathbb{1}(x(T_n) > x^*) \frac{P_n(T_n, x(T_n))}{D_{N+1}(T_n, x(T_n))} \right)$$

$$= D_{N+1}(0) \int_{x^*}^{\infty} \left(\int_{-\infty}^{\infty} \frac{P_n(T_{n+1}, x_{n+1})}{D_{N+1}(T_{n+1}, x_{n+1})} \phi(x_{n+1}|x_n) \, dx_{n+1} \right)$$
$$\times \phi(x_n|x(0)) \, dx_n. \qquad (9.16)$$

The functional form $P_n(T_{n+1})/D_{N+1}(T_{n+1})$ in the last line has already been determined in the previous iteration. Hence, at time T_n we can evaluate (at least numerically) (9.16) for different values of x^*. Let us denote this (numerical) value by $J_n(x^*)$. In the case of the Black formula (9.15), we can solve this expression explicitly to find

$$y_n(T_n, x^*) = y_n(0) \exp\left(-\tfrac{1}{2}\bar{\sigma}_n^2 T_n - \bar{\sigma}_n \sqrt{T_n} \, N^{-1}\left(\frac{J_n(x^*)}{P_n(0)} \right) \right). \qquad (9.17)$$

With the functional form of the swap rate, we can determine the functional form of the numeraire $D_{N+1}(T_n)$ as follows. The swap rate can be written as

$$y_n(T_n) = \frac{1 - D_{N+1}(T_n)}{P_n(T_n)} = \frac{\frac{1}{D_{N+1}(T_n)} - 1}{\frac{P_n(T_n)}{D_{N+1}(T_n)}}. \tag{9.18}$$

Solving for D_{N+1} gives

$$D_{N+1}(T_n, x^*) = \left(1 + y_n(T_n, x^*)\frac{P_n(T_n, x^*)}{D_{N+1}(T_n, x^*)}\right)^{-1}, \tag{9.19}$$

where the term P_n/D_{N+1} has already been calculated in (9.16).

With the procedure outlined above, we can solve the functional forms of the numeraire discount bonds iteratively backwards from time T_{N-1} to time T_1.

9.4 Numerical Implementation

For an implementation of MF models we rely heavily on the evaluation of expectations using numerical integration routines. For a toy model one could use simple numerical integration schemes like the trapezoid rule or Simpson's rule (see Press et al. (1992, Chapter 4)) on a grid of fixed spacing for the Markov process x. This yields reasonable accurate results for pricing. However for the calculation of hedge parameters, we essentially have to take derivatives which implies that the approximation becomes much worse as can be seen by the instabilities in the calculated values of deltas, vegas and especially the gammas.

Another problem with numerical integration on a fixed grid, is that when the calculation date is approaching a fixing date, the probability distribution of the next rate will have very little variance. This leads to integrands which are very spiked, in which case the numerical integration become inaccurate. To make matters worse, as we are working on a fixed grid, all the probability mass of the distribution can become so concentrated that it completely falls in between two grid points. In this case, the numerical integration routine will incorrectly calculate a value of the integral very close to zero.

To overcome these problems we describe several strategies for a successful implementation of a MF model.

9.4.1 Numerical Integration

The numerical integration method we discuss here was introduced by Hunt and Kennedy (2000) and is based on the following idea:

1. fit a polynomial to the payoff function defined on the grid;

2. calculate analytically the integral of the polynomial against the Gaussian distribution.

The only error we now make in the integration is the error introduced by the polynomial fit, since the integration of the polynomial is done analytically. The fitting error can be controlled, by choosing a polynomial of sufficiently high order, and by controlling the spacing in the grid. Further details on the error analysis can be found in Hunt and Kennedy (2000).

Fitting a Polynomial. Given a number of points x_i and a set of function values f_i a polynomial that passes through these values can be computed recursively using Neville's algorithm (see Press et al. (1992, Chapter 3.1)). Let $P_{(i)\cdots(i+m)}$ denote the polynomial defined using the points $x_i, \ldots, x_{i+m}$. Then the following relationship generates higher order polynomials

$$P_{(i)\cdots(i+m)} = \frac{(x - x_{i+m})P_{(i)\cdots(i+m-1)} + (x_i - x)P_{(i+1)\cdots(i+m)}}{x_i - x_{i+m}} \quad (9.20)$$

and $P_{(i)} = f_i$. Each polynomial can be written as $P_{(i)\cdots(i+m)} = \sum_{k=0}^{m} c_{i,k} x^k$. Using (9.20) we can then derive a recurrence formula for the coefficients $c_{i,k}$ as follows

$$c_{i,m} = \frac{c_{i,m-1} - c_{i+1,m-1}}{x_i - x_{i+m}} \quad (9.21)$$

$$c_{i,k} = \frac{x_i c_{i+1,k} - x_{i+m} c_{i,k} + c_{i,k-1} - c_{i+1,k-1}}{x_i - x_{i+m}} \quad \forall k \in [1, m-1] \quad (9.22)$$

$$c_{i,0} = \frac{x_i c_{i+1,0} - x_{i+m} c_{i,0}}{x_i - x_{i+m}}. \quad (9.23)$$

Integrating against Gaussian. The Markov process x defined in (9.1) has Gaussian density functions. Hence, the calculation of integrals against a Gaussian density can be broken down to evaluating for different powers x^k of the polynomial P the following integral

$$G(k; h, \mu, \sigma) = \int_{-\infty}^{h} x^k \frac{\exp\{-\frac{1}{2}(\frac{x-\mu}{\sigma})^2\}}{\sigma\sqrt{2\pi}} dx. \quad (9.24)$$

Using partial integration, we derive the following recurrence relation for G in terms of k

$$G(k) = \mu G(k-1) + (k-1)\sigma^2 G(k-2) - \sigma^2 h^{k-1} \frac{\exp\{-\frac{1}{2}(\frac{h-\mu}{\sigma})^2\}}{\sigma\sqrt{2\pi}}, \quad (9.25)$$

with $G(0) = N(\frac{h-\mu}{\sigma})$ and $G(-1) = 0$.

Calculating Expected Values. Given a grid on which we are working, option values are calculated by taking expectations of the value function against the Gaussian density. Given that we have calculated several option values at time T_{n+1} at grid points x_j, we want to calculate option vales at time T_n for grid points x_i. To do this we proceed as follows:

- Given an order M the approximating polynomial $P_{(j-M/2)\cdots(j+1+M/2)}$ for the interval $[x_j, x_{j+1}]$ is fitted through the points $x_{j-M/2}, \ldots, x_{j+1+M/2}$, where $M/2$ denotes integer division (M div 2).
- We calculate the expectation $\mathbb{E}(f(x, T_{n+1})|x_i)$ given we are currently in point x_i, by adding the integrals of the approximating polynomials against the Gaussian density over all the intervals $[x_j, x_{j+1}]$.
- Loop over all points x_i.

The fitting of the polynomials works well if the function that one wants to approximate is smooth. However, many option payoffs are determined as the maximum of two functions. This implies that the payoff function will be smooth except at the crossover point where the payoff function may have a kink, i.e. a non-differentiable point. Since polynomials are "stiff" they will fit a function with a kink poorly. The way to solve this problem is to fit polynomials to both underlying functions, and to split the integration interval at the crossover point, using the appropriate approximating polynomial on either side of the crossover point.

9.4.2 Non-Parametric Implementation

To determine the functional forms of the numeraire discount bond in the MF model one can use the method described in Sections 9.2 and 9.3. In econometric terms one could call this a *non-parametric* fit of the functional forms. The advantage is that the MF model exactly replicates the market prices of caplets or swaptions. A disadvantage is that this method can be susceptible to numerical errors. In the implementation of the algorithm, the fitted functional forms from time T_{n+1} are the basis for the numerical calculations at time T_n, hence there is a possibility that numerical errors propagate during the algorithm.

The numerical integration algorithm of Hunt and Kennedy described above has very high accuracy and for most implementations of the MF model the accumulation of numerical errors is not an issue. Indeed, all the examples given in this chapter use the non-parametric fitting of the model.

However, there is a tendency in the market to price derivatives with very long maturities (30 or 50 years). When fitting an MF model with 50 year maturity with quarterly time-steps, this means 200 functional forms have to be fitted. Under these circumstances the build-up of numerical errors can be an issue. Furthermore, as we have (at best) only a few different strikes for each maturity where market data is available, imposing that a model exactly fits the Black formula across a continuum of strikes is little more than an assumption on how to interpolate the option prices between various strikes. Hence, we present an alternative implementation in the next subsection.

9.4.3 Semi-Parametric Implementation

In this subsection, we want to propose a different approach for determining the functional forms of the numeraire discount bonds at different maturities. To obtain a better handle on numerical stability we want to use a *semi-parametric* functional form. This means that we fix a functional form with several free parameters that is flexible enough to provide a good fit to the observed market prices. Also, for a suitable choice of the functional form, we can analytically calculate the prices of discount bonds and options on discount bonds thereby eliminating a source of errors in the calibration procedure. The price we will pay is, of course, that we can no longer fit exactly to the Black formula across all strikes.

The functional form we propose for the numeraire discount bond as a function of underlying Markov process x is:

$$\frac{1}{D_{N+1}(t)} = 1 + a_t e^{b_t x(t)} + d_t e^{-\frac{1}{2} c_t (x(t) - m_t)^2}, \tag{9.26}$$

which has (for each time t) the parameters a, b, c, d and m. The choice for this particular functional form is motivated by the fact that we like to have on a large-scale an exponential form (determined by a and b) with a possibility of a local deviation of size d to that form centred at m which dies away depending on c. To obtain reasonable functional forms, we should have $a, b, c > 0$ and d not too large to ensure that D_{N+1} is a monotonic decreasing function in x.

Once the functional form for the numeraire is set, prices for all discount bonds can be calculated as expectations under the martingale measure $\mathbb{Q}^{N+1}$. From (9.3) we obtain for any discount bond D_S that

$$\frac{D_S(t)}{D_{N+1}(t)} = \mathbb{E}^{N+1} \left(\frac{1}{D_{N+1}(S)} \,\middle|\, \mathcal{F}_t \right)$$

$$= \mathbb{E}^{N+1} \left(1 + a_S e^{b_S x(S)} + d_S e^{-\frac{1}{2} c_S (x(S) - m_S)^2} \,\middle|\, x(t) \right). \tag{9.27}$$

If $x(S)|x(t)$ has a Gaussian distribution with mean $x(t)$ and variance $s^2(t, S)$, we can calculate the expectation analytically. This leads to

$$\frac{D_S(t)}{D_{N+1}(t)} = 1 + \left(a_S e^{\frac{1}{2} b_S^2 s^2} \right) e^{b_S x(t)} + \left(\frac{d_S}{\sqrt{1 + c_S s^2}} \right) e^{-\frac{1}{2} \left(\frac{c_S}{1 + c_S s^2} \right) (x(t) - m_S)^2}$$

$$= 1 + a_{S,t} e^{b_S x(t)} + d_{S,t} e^{-\frac{1}{2} c_{S,t} (x(t) - m_S)^2}. \tag{9.28}$$

Note that the functional form for the numeraire rebased discount bonds is preserved under the expectation operator.

Given our analytical formulae for discount bonds, we can also calculate analytically the price of an option on a discount bond. Let $p(T, S, K)$ denote the value at time 0 of a European-style put option with exercise price K and maturity T on a discount bond with maturity S. Of course, we should have $t < T < S$. The price of this option can be calculated under $\mathbb{Q}^{N+1}$ as

$$\frac{p(T, S, K)}{D_{N+1}(0)} = \mathbb{E}^{N+1}\left(\max\left\{K\frac{1}{D_{N+1}(T)} - \frac{D_S(T)}{D_{N+1}(T)}, 0\right\}\right). \tag{9.29}$$

Both the relative discount bond prices have the functional form given in (9.26). The option premium can be evaluated by considering the following result

$$\int_{-\infty}^{h}\left(1 + ae^{bx} + de^{-\frac{1}{2}c(x-m)^2}\right)\phi\left(\frac{x-\mu}{\sigma}\right)dx$$

$$= N\left(\frac{h-\mu}{\sigma}\right) + ae^{b\mu+\frac{1}{2}b^2\sigma^2}N\left(\frac{h-(\mu+b\sigma^2)}{\sigma}\right)$$

$$+ \frac{d}{\sqrt{1+c\sigma^2}}e^{-\frac{1}{2}\frac{c(\mu-m)^2}{1+c\sigma^2}}N\left(\frac{h-\frac{\mu+mc\sigma^2}{1+c\sigma^2}}{\sqrt{\frac{\sigma^2}{1+c\sigma^2}}}\right). \tag{9.30}$$

Note, that to calculate the option premium, we have to apply this result twice.

The price of a call option $c(T, S, K)$ on a discount bond can be expressed in a similar way. Given a pricing formula for options on discount bonds, it is not difficult to derive a pricing formula for caplets or swaptions. We refer to Chapter 5 in the first part of this book for details.

Table 9.1. Semi-parametric Markov-Functional model

$$\Delta T = 1, \alpha = 1, D_n(0) = (1.05)^{-n}$$

T_n	1	2	3	4	5	6	7	8	9
a_n	0.413	0.350	0.285	0.232	0.194	0.160	0.129	0.088	0.045
b_n	0.274	0.269	0.262	0.246	0.226	0.206	0.183	0.166	0.150
c_n	0.022	0.022	0.019	0.015	0.013	0.012	0.016	0.014	0.000
d_n	0.124	0.103	0.096	0.086	0.060	0.037	0.013	0.005	0.000
m_n	0.421	0.515	1.499	2.857	2.931	2.734	0.000	0.000	0.000
$\bar{\sigma}_n$	15%	18%	21%	20%	19%	18%	17%	16%	15%
Err	0.02	0.04	0.08	0.08	0.07	0.06	0.07	0.03	0.00

To fit a semi-parametric MF model to market data, we solve for each time-point T_n for the parameters a_n, b_n, c_n, d_n, m_n such that the error between the model prices and the market prices is minimised. In Table 9.1 we report the results of such a fit for a LIBOR model. We have used annual LIBOR rates with $T_n = n$ and $\alpha \equiv 1$. The initial term-structure of interest rates is given by $D_{T_n} = (1.05)^{-n}$. The numeraire discount bond is D_{10}; hence we have fitted caplet prices at times T_9 back to time T_1.

The row labelled "Err" shows the maximum fitting error in basis-points for the caplet prices at each time T_n. At each time we used the Black formula with volatility $\bar{\sigma}_n$ reported in the table. It is clear from the table that (at least for this example) an accurate fit can be obtained on the basis of the functional form given in (9.26). Note however, that the fitting error seems to increase with the volatility level of the caplet. This may cast some doubt on the applicability of this particular functional form for extreme volatility environments like JPY.

9.5 Forward Volatilities and Auto-Correlation

In this section we present a discussion on mean-reversion and auto-correlation. Then we present a simple choice of the volatility function τ of the Markov process given in (9.1).

9.5.1 Mean-Reversion and Auto-Correlation

Mean-reversion of interest rates is considered a desirable property of a model because it is perceived that interest rates tend to trade within a fairly tightly defined range. This is indeed true, but when pricing exotic derivatives it is the effect of mean-reversion on the auto-correlation of interest rates that is more important. By *auto-correlation* we mean the correlation of a process at different point in time. This in contrast to *cross-correlation* which is the correlation of different processes observed at the same point in time. For one-factor models the cross-correlation is always equal to 1, but this does not mean we have no control over the auto-correlation. We illustrate this for the Hull-White model because it is particularly tractable.

In the Hull-White model (which was analysed in Chapter 5) the short-rate process $r(t)$ follows

$$dr = (\theta(t) - ar)dt + \sigma\, dW. \tag{9.31}$$

The general auto-correlation structure of the spot interest rates depends on the mean-reversion parameter a via the relationship

$$\mathrm{Corr}\big(r(t), r(s)\big) = \sqrt{\frac{e^{2at} - 1}{e^{2as} - 1}}, \tag{9.32}$$

for $t < s$. Thus, increasing the mean-reversion a has the effect of reducing the auto-correlation between the short rate at different times.

9.5.2 Auto-Correlation and the Volatility Function

The question is now, how can we endow the MF model with mean-reversion thereby placing the auto-correlation of the interest rates under our control.

We have parametrised our MF example in terms of the Markov process x. The auto-correlation between the interest rates in the MF model is governed by the auto-correlations of the process x. If we set $\tau(t) = e^{at}$ in equation (9.1) we find for the auto-correlation of the process x (for $t < s$)

$$\mathrm{Corr}\big(x(t), x(s)\big) = \sqrt{\frac{e^{2at} - 1}{e^{2as} - 1}}, \qquad (9.33)$$

which is the same correlation structure as in the Hull-White model. Hence, we can interpret the parameter a as the mean-reversion parameter of the MF model. For $a = 0$ (9.33) reduces to $\sqrt{t/s}$ which is the auto-correlation structure of Brownian motion. Because the fitted functional forms are non-linear functions of the x process, the auto-correlation between the interest rates will only approximately follow the relation (9.33). What is important however, is not so much the exact functional form, but the fact that even though the MF models presented here are one-factor models, we can control the auto-correlation between the interest rates. As we will see in the examples later on, these correlations have a crucial impact on the valuation of exotic interest rate contracts.

9.6 LIBOR Example: Barrier Caps

In this section we revisit the example of barrier caps of Chapter 8. Barrier options are products that can be valued both with Monte Carlo simulation in the LIBOR market model or on a grid in a LIBOR MF model; hence this product enables us to make a comparison between both valuation methodologies.

9.6.1 Numerical Calculation

To explain how we proceed numerically to value (double barrier) discrete barrier caps, in the MF model we turn to Figure 9.1. In this figure, the vertical lines represent different times T_i. The dots on these lines are the values for the x-process where explicit (numerical) calculations are performed. The intermediate values are obtained by fitting polynomials. As usual calculation proceeds backward, that is from right to left in the figure. Asset values (discount bond prices and option prices) are obtained by taking expectations under the martingale measure $\mathbb{Q}^{N+1}$.

The expectations are obtained numerically by integrating the fitted polynomials analytically against the Gaussian density functions of the x process (as explained in Section 9.5.1).

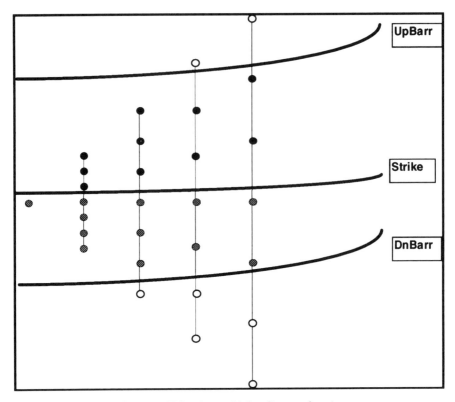

Fig. 9.1. Valuation grid for discrete barrier

The grid contains four types of nodes:

- knocked out at the down-barrier (white nodes)
- not knocked out and out-of-the-money (grey nodes)
- not knocked out and in-the-money (black nodes)
- knocked out at the up-barrier (white nodes)

This is in case of a call option, for put options the grey and black nodes change roles.

The option values are calculated backward, say from T_{i+1} to T_i. For every node at time T_i the option values is calculated by taking the expected value of the option at time T_{i+1} and adding the current payoff. The integral for taking the expected value is divided into 4 pieces. Suppose the points x_{dn}, x_{str} and x_{up} are the values of the x process for which $L_{i+1}(T_{i+1})$ is equal to the down-, strike and up-barrier respectively. These values are determined by finding the x values for which the polynomial fitted through the LIBOR rates is equal to these levels. Below x_{dn} the option is knocked out and the value from later caplets is nullified. Between x_{dn} and x_{str} (grey nodes) the option is not knocked out, but is out-of-the-money and no new payoff is received but the value from later caplets is retained. Between x_{str} and x_{up} (black nodes) the option is not knocked out and is in-the-money and a new payoff is received in addition to the value from later caplets. Above x_{up} the option is also knocked out and the value from later caplets is nullified.

9.6.2 Comparison with LIBOR Market Model

In Table 9.2 we compare the prices calculated by the MF model with the prices calculated by the LIBOR market model as reported in Table 8.3. We denote the price in basispoints obtained by simulation in the LIBOR market model by "LMM" and the standard error of the simulation by "SE". The prices of the Markov-Functional model with a mean-reversion $a = 0\%$ are denoted by "MF".

Table 9.2. Prices of discrete barrier caps/floors

Semi-Annual, $M = 100,000$, Flat Vol 10.00%, MeanRev 0%

Up & Out Cap					
Mat	Strike	Barrier	LMM	(SE)	MF
2Y	4.00%	5.00%	37.69	(0.08)	37.60
2Y	4.00%	7.00%	196.80	(0.22)	196.70
3Y	4.50%	5.50%	39.60	(0.09)	39.65
3Y	4.50%	7.50%	247.40	(0.31)	247.22
5Y	5.00%	6.00%	44.04	(0.11)	44.17
5Y	5.00%	8.00%	334.11	(0.48)	333.78
7Y	5.50%	6.50%	45.92	(0.12)	46.09
7Y	5.50%	8.50%	366.84	(0.61)	366.37
10Y	6.00%	7.00%	46.79	(0.13)	46.69
10Y	6.00%	9.00%	387.00	(0.74)	386.82
Up & Out Floor					
2Y	4.00%	3.00%	0.00	(0.00)	0.00
2Y	4.00%	5.00%	1.57	(0.01)	1.58
2Y	4.00%	7.00%	1.57	(0.01)	1.58
3Y	4.50%	3.50%	0.16	(0.01)	0.16
3Y	4.50%	5.50%	16.06	(0.06)	16.10
3Y	4.50%	7.50%	16.06	(0.06)	16.10
5Y	5.00%	4.00%	11.77	(0.08)	11.84
5Y	5.00%	6.00%	52.47	(0.13)	52.47
5Y	5.00%	8.00%	52.51	(0.13)	52.51
7Y	5.50%	4.50%	65.32	(0.18)	65.43
7Y	5.50%	6.50%	114.80	(0.27)	114.76
7Y	5.50%	8.50%	115.25	(0.27)	115.20
10Y	6.00%	5.00%	143.48	(0.35)	143.50
10Y	6.00%	7.00%	215.86	(0.55)	216.01
10Y	6.00%	9.00%	219.38	(0.55)	219.46

From the table it is clear that there is very little difference between the values calculated by the MF model and the LIBOR market model. The MF model has the advantage that the calculation time of the MF model is much shorter than the execution time of a Monte Carlo simulation.

9.6.3 Impact of Mean-Reversion

In Table 9.3, the prices (in basis-points) for knock-out caps and knock-out digital floors are compared for different levels of mean-reversion. The valuations were carried out for DEM yield-curves and volatilities observed August 20, 1998. The payoff and knock-out is determined by 6M LIBOR . The columns "MFx%" denote a mean-reversion of x%.

Table 9.3. Impact of mean-reversion

DEM curves on 20aug99, first LIBOR fixing: 24feb99
semi-annual LIBOR rates, option strike=4%

	Up Barr=4.5%			Dn Barr=3.75%		
Cap	MF0%	MF10%	MF20%	MF0%	MF10%	MF20%
Mat=1yr	1.8	1.8	1.8	2.2	2.2	2.2
Mat=5yr	13.8	13.4	13.1	156.7	148.3	139.8
Mat=10yr	16.9	15.9	15.1	419.4	374.3	331.1
Mat=20yr	18.3	16.7	15.6	819.4	658.5	526.3
1% Dig Flr						
Mat=1yr	37.0	37.0	37.0	14.4	14.4	14.4
Mat=5yr	171.3	167.8	164.1	25.9	26.6	26.2
Mat=10yr	235.9	219.0	203.5	28.0	28.3	28.5
Mat=20yr	303.3	252.2	218.4	29.5	30.0	30.0

We see from the table that for short maturities the mean-reversion has relatively little impact on the value of discrete barrier options. For long maturities, the impact can be significant. Furthermore, we see for the knock-out cap and the up and out digital floor that the value decreases as the mean-reversion increases. This can be explained as follows. An increase in mean-reversion implies a decrease in correlation between the forward LIBOR rates. A lower correlation implies that LIBOR rates have less tendency to move in parallel, which implies if today's LIBOR is fixed far away from the barrier there will be more probability that the next fixing will be close to the barrier, therefore increasing the probability of knock-out and decreasing the value of the option.

9.7 LIBOR Example: Chooser- and Auto-Caps

In this section we present an analysis of chooser- and auto-caps. For further details we refer to Pedersen and Sidenius (1997).

9.7.1 Auto-Caps/Floors

An *auto-cap/floor* (also known as *limit cap/floor*) comprises a total of M caplets along with an associated limit m. The customer receives a payment equal to the payment of a normal caplet. However, only the first m caplets that fix in-the-money will be paid to the customer. In the case that $m = M$, an auto-cap/floor reduces to a regular cap.

9.7.2 Chooser-Caps/Floors

Closely related to the auto-cap is the *chooser-cap* (also known as *flexible cap*). There is once again a limit m on the total number of caplets that can be exercised. The difference in this case is that the holder of the option has the right, on each reset date, to decide whether or not to exercise that particular caplet and count it as part of all the m. This decision, once made, cannot be changed at a later date. One possible exercise strategy is to exercise a caplet as soon as it sets in the money, which would replicate the payoff of an auto-cap/floor. Hence, the value of a chooser-cap/floor is always greater or equal to the value of an auto-cap/floor.

Again, in the case that $m = M$, a chooser-cap/floor reduces to a regular cap/floor. In the case that $m = 1$ a chooser can be seen as a Bermudan-style cap since the holder of the option has the right to decide (only once) when to exercise the option.

9.7.3 Auto- and Chooser-Digitals

The definitions given above can easily be extended to include digitals. Note, however, that for digitals there is no distinction between auto- and chooser-digitals. It is easy to see that for chooser-digitals it is always optimal for the holder to exercise the digital caplet as soon as it is set in-the-money. There is no value added in waiting since the payoff will always be the same, only the discounting will be higher (and we have the probability that none of the future caplets will end in-the-money).

9.7.4 Numerical Implementation

Auto- and chooser-caps are path-dependent instruments. However, the path-dependency can be resolved in a very simple way, by introducing one additional state-variable: the number of caplets left in the contract for exercising.

Table 9.4. Prices of auto- and chooser-caps/floors

Semi-Annual, MeanRev 0%, Strike=5%

$M = 10$	Cap		Floor	
m	Auto	Chooser	Auto	Chooser
1	23.19	122.13	36.71	36.89
2	75.14	238.43	54.89	56.01
3	150.09	347.77	69.48	71.46
4	241.47	448.21	83.36	85.37
5	343.29	538.20	95.99	97.72
6	448.62	615.38	107.58	108.70
7	551.86	675.52	117.16	117.84
8	646.76	713.35	124.97	125.35
9	716.98	727.86	131.46	131.60
10	728.86	728.87	136.43	136.43

Denote this state-variable by m. Then $V(m, T_n, x_i)$ denotes the value at time T_n in node x_i of an instrument with m caplets left to exercise. Of course, we have that m is less than or equal to the number of caplets M in a regular cap.

In every node $[T_n, x_i]$ we first calculate the value of the derivative with no early exercise using for every m. The value for an auto- and chooser-caplet are then calculated as follows:

Auto. If the current caplet is in-the-money, then a caplet is exercised automatically and we get

$$V(m, T_n, x_i) = \text{Payoff}(T_n, x_i) + V(m - 1, T_n, x_i). \qquad (9.34)$$

If $m = 0$, then we have no caplets left to exercise and no payoff gets added. Also, if the current caplet is out-of-the-money, no payoff is added.

Chooser. For a chooser-cap, the holder of the option has the right to make the optimal decision whether the current value of early exercising is larger than the loss of flexibility by loosing one caplet. Hence, we update the chooser-caplet as follows

$$V(m, T_n, x_i) = \max\{V(m, T_n, x_i), \text{Payoff}(T_n, x_i) + V(m - 1, T_n, x_i)\}. \qquad (9.35)$$

Again, if m=0 then there are no caplets left and no payoff gets added.

In Table 9.4 we compare prices for auto- and chooser-caps and floors. For the calculations we used a yield-curve given by the zero curve $Z(T) = 0.08 - 0.05 \exp\{-0.18T\}$. The prices of the discount bonds at time 0 are thus given by $D_n(0) = \exp\{-Z(T_n)T_n\}$. To calculate the LIBOR fixing dates T_n we made the simplifying assumption that $T_n = 0.5n$ for semi-annual payments, with daycount fractions α_n being constant and equal to 0.5. Furthermore we used to following term-structure of caplet volatilities for $\bar{\sigma}_1$ until $\bar{\sigma}_{10}$: 16.50%, 16.50%, 19.00%, 21.50%, 22.50%, 23.50%, 22.63%, 21.75%, 21.75%, 21.75%.

From the table we see that the prices of chooser-caps are much higher than the prices of auto-caps. This is to be expected, since we have an increasing term-structure of interest rates. When a caplet gets exercised in an auto-cap, it is probably not very far in-the-money. However, in a chooser-cap it may be advantageous to wait until a later date and exercise the caplet when it is much deeper in-the-money. Indeed, for the auto-floor and chooser-floor, we see relatively little difference in value. In an upward-sloping yield-curve, floorlets with longer maturities are more out-of-the-money. Hence, the optimal strategy of the chooser-floor is closely related to the exercise strategy of the auto-floor which is to exercise a floorlet as soon as it sets in-the-money.

9.8 Swap Example: Bermudan Swaptions

The products we want to consider in this section are Bermudan-style swaptions. These are options which give the right, on a given set of dates, to exercise into an underlying swap. The algorithm we describe here is designed to price Bermudans of the "cancellable swap" type. This means that the end-date of all the underlying swaps in one particular deal is the same and furthermore that the cash flows for the overlapping parts of the underlying swaps are identical.

Once the marginal distributions for the numeraire discount bond have been determined (as described in Section 9.3), we can calculate prices of derivatives using the standard backward induction. During our calculations, we keep track of the numeraire rebased value of the derivative (that is, $V(t)/D_{N+1}(t)$) calculating the (relative) value of intermediate payments and early exercise decisions as required. At time 0, we calculate the real price by multiplying by $D_{N+1}(0)$.

9.8.1 Early Notification

One difficult feature of Bermudan swaptions in practice is *early notification*. This means that the holder of the option has to announce several days in advance (typically 10 or 30 business days) when he wants to exercise the option. Once the announcement is made, the option holder is obliged to exercise, even if the market conditions have turned against him. Clearly, this reduces the value of the option.

To analyse the effect of early notification, we consider a simple example: a European option with early notification. Let $0 < t < T$, where T is the maturity of the option and t is the notification date. What is the value of this option at time 0? At time t the holder of the option has to decide whether he wants to exercise his option or not. The option holder has to determine an optimal value K^* to announce or not. If he decides to make the announcement, he has a forward position for the interval $[t, T]$; if he does not exercise, his option value will be zero. At time t his payoff is given by $V(t) = D_T(t)[F(t)-K]\mathbb{1}_{F(t)>K_*}$, where $F(t)$ denotes the forward value of the underlying. The value at time 0 is given by $V(0)/D_T(0) = \mathbb{E}^T[V(t)/D_T(t)]$. The optimal choice of K^* is the value that maximises $V(0)$. It is easy to see that this choice is given by $K^* = K$. Hence, the optimal point for early notification is given by making the normal exercise decision but based on the forward price. In the MF model we therefore implement early notification by making the normal early exercise decision based on the forward swap rates at the notification date.

9.8.2 Comparison Between Models

For a 30 year DEM Bermudan, which is exercisable every five years we have compared in Table 9.5 below, for different levels of mean-reversion, the prices calculated by three different models: Black and Karasinski (1991) (BK), MF and the generalised Hull-White model (HW). (The table and subsequent analysis presented here can also be found in Hunt, Kennedy and Pelsser (2000).)

For every level of mean-reversion[15], we have given the prices of the embedded European swaptions (5×25, 10×20, 15×15, 20×10, 25×5). Since all three models are calibrated to these prices, all models should agree exactly on these prices. The differences reported in the table are due to numerical errors. We adopted the following approach. First we chose a level for the BK mean-reversion parameter and reasonable levels for the BK volatilities. We then used these parameters to generate the prices, using the BK model, of the underlying European swaptions. We then calibrated the other, HW and MF, models to these prices. The resultant implied volatilities used for each case are reported in the second column. At the bottom of each block in the table, we report the value of the Bermudan swaption as calculated by each of the three models.

From the table we see that the mean-reversion parameter has a significant impact on the price of the Bermudan swaption. It is intuitively clear why this is the case. The reason a Bermudan option has more value than the maximum of the embedded European option prices is the freedom it offers to delay or advance the exercise decision of the underlying swap during the life of the contract to a date when it is most profitable. The relative value between exercising "now" or "later" depends very much on the correlation of the underlying swap-rates between different time points. This correlation structure is exactly what is being controlled by the mean-reversion parameter. By contrast, the effect of changing the model is considerably less, and what difference there is will be due in part to the fact that the mean-reversion parameter has a slightly different meaning for each model. We conclude that

---

[15]The calculations on the HW model were done on an implementation that does not allow negative mean-reversion; hence no values for the HW model are reported in this case.

Table 9.5. Comparison of Bermudan swaption prices

Strike: 0.0624, Currency: DEM, Valuation date: 11-Feb-98

Mean-reversion = -0.05							
European		Receiver			Payer		
Mat	ImVol	BK	MF	HW	BK	MF	HW
05 × 25	8.17%	447.7	445.9	–	456.6	456.6	–
10 × 20	7.74%	365.2	363.8	–	448.2	448.6	–
15 × 15	7.90%	285.8	284.7	–	350.8	351.4	–
20 × 10	8.29%	191.8	190.8	–	241.6	242.1	–
25 × 05	8.68%	91.0	90.9	–	126.2	126.1	–
Bermudan		510.0	502.7	–	572.2	566.9	–
Mean-Reversion = 0.06							
European		Receiver			Payer		
Mat	ImVol	BK	MF	HW	BK	MF	HW
05 × 25	8.46%	463.7	462.8	464.2	472.6	472.0	473.3
10 × 20	7.91%	374.0	373.5	374.2	457.0	456.9	457.5
15 × 15	7.80%	281.8	281.5	281.9	346.8	346.8	347.1
20 × 10	7.81%	179.5	179.4	179.6	229.4	229.3	229.5
25 × 05	8.16%	84.8	84.7	84.7	119.9	120.0	120.1
Bermudan		606.1	602.9	608.2	743.6	717.7	727.3
Mean-Reversion = 0.20							
European		Receiver			Payer		
Mat	ImVol	BK	MF	HW	BK	MF	HW
05 × 25	8.37%	458.7	457.6	459.3	467.5	466.8	468.0
10 × 20	6.99%	326.2	325.4	326.2	409.1	408.8	409.3
15 × 15	6.67%	237.1	236.4	237.3	302.1	301.7	302.4
20 × 10	7.32%	166.7	166.4	166.6	216.6	216.4	216.6
25 × 05	9.40%	99.8	99.5	99.6	134.9	134.8	134.8
Bermudan		647.4	665.8	673.9	885.5	814.7	813.6

the precise marginal distributional assumptions made have a secondary role
in determining prices for exotic options relative to the joint distributions as
captured by the mean-reversion parameter.

10. An Empirical Comparison of Market Models

The market models presented in Chapter 8 and the Markov-Functional models presented in Chapter 9 have obvious advantages for pricing exotic interest rate derivatives over the spot interest rate models presented in Part I of this book. It is therefore not surprising these models have received a lot of attention recently. These has been relatively little attention however for the empirical performance of these models. In this chapter we summarise some results from a recent study by De Jong, Driessen and Pelsser (1999) (DJDP hereafter) on the empirical performance of LIBOR market models (LMM) and swap market models (SMM). For a more detailed account, including a rigorous econometric analysis of the models, the interested reader is referred to the De Jong-Driessen-Pelsser paper.

In this chapter we focus on two important results from their study. The first is a direct comparison of the LMM and the SMM. Since the LMM and SMM are mutually inconsistent, it is an empirical question which model is to be preferred for practical purposes. The way DJDP analyse the question is to fit the LMM to the market prices of caplets and then calculate the implied prices of swaptions and compare these prices to prices of swaptions quoted in the market. On the other hand, DJDP fit a SMM to a set of swaption prices and then calculate the implied prices of caplets and compare those to the market prices. The empirical results show that the LMM in general leads to better prediction of the swaptions prices than the SMM which tends to substantially overprice caplets. An explanation for this phenomenon is provided.

Second, DJDP investigate the specification of the volatility functions. The importance of this issue has already been highlighted in Chapter 9. DJDP show that the usual choice of a constant volatility function (mean reversion equal to 0) is not a particularly good one. The LMM with a constant volatility function persistently overprices swaptions and the SMM with a constant volatility function underprices swaptions. Much better results are obtained by specifying an increasing volatility function (which corresponds to positive mean-reversion).

The rest of this chapter is organised as follows. In Section 10.1 we describe the data used in the DJDP study. In Section 10.2 first discusses the calibration methodology for the LMM and then presents the results, Section 10.3 presents the calibration methodology and results for the SMM. Finally, we conclude in Section 10.4.

10.1 Data Description

DJDP use two datasets for their analysis: USD term structure data and USD implied volatility data for caplets and swaptions.

The term structure data consist of daily observations on a spot LIBOR rate, Eurodollar futures prices and par swap rates of different maturities. These observations are available for the period July 1995 to September 1996. The prices of Eurodollar futures contracts do not give directly information about the forward LIBOR rates, but a futures/forward correction has to be applied first, DJDP use the correction derived by Brace (1998). The term-structure of interest rates is calculated using the assumption that the forward LIBOR rates are a piecewise linear function of the maturity.

The second dataset consists of quotes for implied Black volatilities for at-the-money 3-months caplets for different maturities and at-the-money swaptions with different option maturities and swap tenors. These observations are also available for the period July 1995 to September 1996: 282 trading days in total. The caplet maturities range from 3 months to 10 years, the option maturities for the swaptions range from 1 month to 5 years and the swap tenors range from 1 to 10 years.

10.2 LIBOR Market Model

In this section we analyse the LIBOR market model. We start with a description of the calibration methodology and the different choices made for the volatility function. Then we present the pricing results for caplets and swaptions found by DJDP

10.2.1 Calibration Methodology

DJDP base the analysis of the models on three complementary sets of derivatives:

- caplets;
- < 10 year total maturity swaptions;
- 10 year total maturity swaptions.

The total maturity is defined as the sum of the option maturity and the swap tenor, i.e. the last date on which a payment is made in the underlying swap.

The empirical setup is such that the LMM is calibrated using the observed caplet implied volatilities, whereas the SMM is calibrated to the 10 year total maturity swaptions. Both models are implemented using the 10 year time point as the terminal time. Hence, the pricing errors of the LMM and SMM for the < 10-swaptions can be used to compare the accuracy of the models.

DJDP calibrate a one-factor LMM to the implied volatilities of caplets and analyse what the implications of this model are for the prices of swaptions. A simple choice for the volatility function σ_n of LIBOR rate L_n is to set

$$\sigma_n(t) = \sigma_n = \bar{\sigma}_n, \tag{10.1}$$

where $\bar{\sigma}_n$ denotes the implied Black volatility of the caplet with maturity T_n.

The LMM can be endowed with mean reverting behaviour by choosing an increasing volatility function. A simple choice for an increasing volatility function is given by

$$\sigma_n(t) = \sigma_n e^{at}. \tag{10.2}$$

To fit a LMM to the market prices we have to solve

$$\int_0^{T_n} \sigma_n^2 e^{2as} \, ds = \bar{\sigma}_n^2 T_n \tag{10.3}$$

for σ_n which yields

$$\sigma_n = \bar{\sigma}_n \sqrt{\frac{2aT_n}{e^{2aT_n} - 1}}. \tag{10.4}$$

We will refer to the LMM without mean-reversion as the *constant volatility LMM*, as the volatility function is constant for each LIBOR rate. The LMM with mean-reversion will be called *mean-reversion LMM*. Although the constant volatility LMM is a special case of the mean-reversion LMM (with $a = 0$) we do consider the constant volatility LMM as a special case as most empirical work on market models to date (for example Longstaff, Santa-Clara and Schwartz (1999)) has only considered constant volatility specifications of the LMM and has ignored mean-reversion.

For the mean-reversion LMM, the mean-reversion parameter cannot be determine from the market prices of caplets, but has to be estimated otherwise. In the literature, mean-reversion estimates are often obtained by a time-series analysis of interest rates (Chan et al. (1992)), by fitting a cross-section of bond prices (De Munnik and Schotman (1994)) or both (De Jong (1999)). DJDP follow a different approach. As the goal is to accurately price swaptions with the LMM, it is natural to estimate the mean-reversion parameter a from the daily cross-section of observed swaptions prices. More specifically, DJDP estimate the mean-reversion parameter each day by minimising the sum of squared differences between the observed volatilities for all swaptions and the implied volatilities of the prices generated by the LMM.

Table 10.1. Pricing errors for LIBOR market model

Errors measured in volatility percentage points.

Mean Reversion	Average Error	Average AbsErr	Average MaxErr
Swaptions < 10 year			
No	1.84	2.11	4.41
Yes	0.30	1.08	3.88
Swaptions = 10 year			
No	2.01	2.08	2.99
Yes	-0.80	0.90	1.93

10.2.2 Estimation and Pricing Results

In Table 10.1, we give an overview of the pricing errors of the swaption prices for the LMM measured in terms of implied volatilities found by DJDP. We have reported the average of all pricing errors over the sample of 282 daily observations. We have also reported the average of the absolute pricing errors ("AbsErr") and the average of the maximum error on each day ("MaxErr").

Let us first consider the constant volatility LMM. For this model the average pricing errors of the swaptions are positive. This indicates that swaptions are *overpriced* by the constant volatility LMM. A standard explanation for this result, put forward by Longstaff, Santa-Clara and Schwartz (1999) and Rebonato (1998), is a missing second factor. A second factor induces additional decorrelation between the forward LIBOR rates which in turn leads to lower swaption volatilities.

An alternative explanation for the overpricing of swaptions, which we feel has been ignored in the literature, is the absence of mean reversion. As explained in Chapter 9, mean-reversion also leads to additional decorrelation between the LIBOR rates which leads to lower swaption volatilities.

To investigate the effect of mean-reversion, DJDP have estimated the mean-reversion LMM. Given the fact that the mean-reversion parameter is obtained by a fit to swaption prices, it is clear that the mean-reversion LMM always has a better fit than the constant volatility LMM. However, as shown in Table 10.1, the increase in the fit of swaption prices is large. With the addition of only one parameter, a reduction in the average of the absolute errors of around 50% can be obtained. Also DJDP find that the estimated values of the mean-reversion parameter are quite stable over time, which is an indication that the mean-reversion parameter is not merely fitted to noise in the data.

10.3 Swap Market Model

DJDP perform a similar analysis for the SMM and compare the results for the LMM and SMM.

10.3.1 Calibration Methodology

For the analysis DJDP calibrate a SMM to the set of swaptions with a total maturity equal to 10 years. With this choice, they can then price all the caplets and all swaptions with a total maturity less than 10 years. Like in the LMM we can specify a *constant volatility SMM* that has the volatility function as a constant for each forward swap rate y_n, like in (10.1). Or we can construct a *mean-reversion SMM* by choosing the volatility functions like in (10.2).

For the mean-reversion SMM, the mean-reversion parameter a is estimated daily by minimising the sum of squared volatility errors over all instruments not priced directly by the SMM, i.e. swaptions with a total maturity less than 10 years and all caplets. The prices of these caplets and swaptions cannot be determined analytically for the SMM and thus we use Monte Carlo simulation to price these instruments.

Table 10.2. Pricing errors for swap market model

Errors measured in volatility percentage points

Mean Reversion	Average Error	Average AbsErr	Average MaxErr
Caplets			
No	2.87	6.99	26.11
Yes	5.16	7.36	28.09
Swaptions < 10 year			
No	-1.22	2.26	12.05
Yes	0.32	1.77	12.48

10.3.2 Estimation and Pricing Results

In Table 10.2 we give an overview of the pricing errors of the SMM measured in terms of implied volatilities found by DJDP. We have reported the average of all pricing errors over the sample of 282 daily observations. We have also reported the average of the absolute pricing errors ("AbsErr") and the average of the maximum error on each day ("MaxErr").

Most remarkable feature from this table are the large pricing errors for caplets generated by the SMM. Once the SMM is calibrated to the prices of swaptions, it does a very poor job pricing caplets. This effect is persistent whether or not mean-reversion is included.

An intuitive explanation for this phenomenon can be given as follows. In a swap contract various floating and fixed payments are exchanged. As the floating payments are based on LIBOR rates, we can view the a swap contract as a portfolio of forward LIBOR payments. Hence, the volatility of the forward par swap rate is determined by the volatility of the forward LIBOR rates. To first order, we can say that the volatility of a forward par swap rate is the volatility of a weighted average of forward LIBOR rates.

If we are given a set of LIBOR volatilities, we can impute the swap rate volatilities. Due to the averaging, individual differences in the observed LIBOR volatilities are cancelled out. If, on the other hand, we are given a set of swaption volatilities we can reverse the calculation and impute the individual LIBOR volatilities. However, small differences in the swaption volatilities (due to measurement errors like bid/ask spreads in the data) tend to get magnified when calculating the LIBOR volatilities. It is exactly this effect we see in Table 10.2.

10.4 Conclusion

DJDP have shown that a constant volatility LMM leads to overpricing of swaptions. Including a mean-reversion term in the volatility specification significantly increases the fit of the LMM. For a mean reversion LMM the average absolute pricing error is around 1 volatility percent point, which is similar in size to the bid/ask spread typically seen in the market. For a SMM we have seen that the model severely misprices caplets. Hence, it seems that a LMM with mean-reversion can reasonably replicate prices of both caplets and swaptions.

However, a warning is in place here. We have presented a very simple analysis here, by only looking on a daily basis at a cross-section of caplet and swaption prices. For a more elaborate study, including rigorous econometric tests on the models and an analysis of two-factor models we refer to De Jong, Driessen and Pelsser (1999).

In reality, the relation between caplet and swaption volatilities is determined by two effects: the auto-correlation of interest rates (as captured by mean-reversion) and the cross-correlation of different interest rates (as captured by a multi-factor model). We have only focussed on mean-reversion to highlight the importance of taking mean-reversion into account, which we feel has been largely ignored in the literature up until now. For a realistic model that is capable of describing the dynamics of LIBOR and swap rates simultaneously we would need to identify the number of factors and the mean-reversion effects simultaneously. Given a data-set like the one in this study, we can pool the time-series and cross-sectional information to estimate these effects simultaneously. However, much more work along these lines is needed to obtain a better understanding of the relationship between swaption and caplet volatilities.

11. Convexity Correction

The general models developed in Chapters 8 and 9 are models that describe the evolution of the whole yield-curve.[16] Hence, these models are quite complicated since one needs to keep track of many stochastic processes (e.g. all LIBOR rates) simultaneously. For the example products described in those chapters, such a level of complexity is indeed required.

Many products that are actively traded in interest rate derivative markets however are much less complicated. The payoff of these products only depends on a few interest rates which are only observed at one point in time. One could characterise such products as *exotic European* options. The value of these products is determined by the (joint) probability distribution of the relevant rates at this one point in time. Hence, we only need to model the joint distribution for the rates under consideration thereby focussing the modelling effort on the problem at hand. The marginal distribution for the individual rates can be inferred from the prices of vanilla products like caps and swaptions. Given the marginal distributions of, say, two rates, the joint distribution can then be parametrised by a correlation parameter. This correlation parameter then captures the additional risk factor of this particular product.

Of course, it is also possible to calculate prices for exotic European products using any of the models discussed in Chapters 8 or 9. These models also fit the marginal distributions of a set LIBOR or swap rates. In the one-factor models, the joint distribution of different rates observed at the same point in time is then determined by the assumptions of the model. It may seem natural to implement a multi-factor model to solve this problem. But to implement a multi-factor model one has to specify the joint probability distribution of the driving factors of the model. This is not a trivial procedure. Often a statistical procedure like principal component analysis is suggested (see for example Heath, Jarrow and Morton (1996)) to extract the factors from historical data. However, as is pointed out by Rebonato (1998, Chapter 3; 1999), using factors based on principal components can lead to very unrealistic correlation structures for particular rates. The danger is very real that an uncritical user may never detect these unrealistic correlations, as they are hidden very deep inside the model.

---

[16]This chapter is based on Pelsser (2000).

We therefore very much advocate the practice of trying to focus the modelling effort as close as possible to the problem at hand, instead of relying on a "big" multi-factor model which may have a lot of hidden assumptions.

In the sections that follow, we consider a broad class of exotic European options that are characterised by the fact that certain interest rates are paid at the "wrong" time and/or in the "wrong" currency. It turns out that the forward interest rate has to be adjusted to reflect the fact the interest rate is paid "incorrectly". This adjustment is known in the market as *convexity correction*. See for example, Hull (2000, Chapter 19 and 20) for an overview. In this chapter, we will put convexity correction on a firm mathematical basis by showing that it can be interpreted as the side-effect of a change of numeraire. This means that we will show how convexity correction can be understood mathematically as the expected value of an interest rate under a different probability measure than its own martingale measure.

The remainder of this chapter is organised as follows. In Section 11.1 we show how convexity correction can be understood as the expected value of an interest rate under a different probability measure. We derive this result both for single currency and for multi-currency economies. In Section 11.2 we show how options on convexity corrected rates can be valued, and we derive a convenient approximation in terms of the Black formula. In Sections 11.3 and 11.4 we give examples of single and multi-index products that can be valued with this approach. We conclude this chapter with a warning in Section 11.5 on the use of convexity correction for products with very long maturities. Section 11.6 is an appendix where we derive the Linear Swap Rate Model on which many of the formulæ in this chapter are based.

11.1 Convexity Correction and Change of Numeraire

In this section we provide an overview of how convexity correction can be understood mathematically as the expected value of an interest rate under a different probability measure than the martingale measure.

First we review how, in an arbitrage-free economy, different probability measures can be used to value a given product. From this we derive the well known Change of Numeraire Theorem. Then we extend this theorem to multi-currency economies. With these tools we are then in a position to put the concept of convexity correction on a firm mathematical basis.

11.1.1 Multi-Currency Change of Numeraire Theorem

In Chapter 2 we have already encountered the Change of Numeraire Theorem (see equations (2.11) and (2.12)). This theorem shows how in an arbitrage-free economy an expectation under a probability measure $\mathbb{Q}^N$ generated by a numeraire N can be represented as an expectation under a probability measure $\mathbb{Q}^M$ generated by a numeraire M times the Radon-Nikodym derivative $d\mathbb{Q}^N/d\mathbb{Q}^M$ which is equal to the ratio of the numeraires N/M. For an expectation at time 0 of a random variable $H(T)$ equation (2.11) reduces to

$$\mathbb{E}^N\left(H(T)\right) = \mathbb{E}^M\left(H(T)\frac{N(T)/N(0)}{M(T)/M(0)}\right), \tag{11.1}$$

where $\mathbb{E}^N, \mathbb{E}^M$ denotes expectation under the probability measure $\mathbb{Q}^N, \mathbb{Q}^M$ respectively.

Many of the products we are interested in will be multi currency products. Hence, we will extend in this section the Change of Numeraire Theorem to include multi currency economies. We will proceed along the same lines as in Chapter 2.

Suppose we have a domestic economy d and a foreign economy f together with the exchange rate $X^{(d/f)}(t)$ that expresses the value at time t of one unit of foreign currency in terms of domestic currency. This immediately implies the relation $X^{(f/d)} = 1/X^{(d/f)}$. Also we assume that this system of economies and exchange rates is arbitrage-free and complete. This implies that for a numeraire $N^{(d)}$ in the domestic economy there exists a unique martingale measure $\mathbb{Q}^{N,d}$ such that all $N^{(d)}$ rebased traded assets in the domestic economy become martingales. Note that we have two types of traded assets in the domestic economy: the domestic assets $Z^{(d)}$ and the domestic value of the foreign assets which are given by $X^{(d/f)}Z^{(f)}$. All these assets are traded assets in the domestic economy and can be used as numeraires if their values are strictly positive.

Let us now consider two numeraires, one in the domestic economy, say $N^{(d)}$, and one in the foreign economy, say $M^{(f)}$. As the exchange rate is strictly positive, the domestic value of the foreign numeraire $X^{(d/f)}M^{(f)}$ is a valid numeraire in the domestic economy. Hence, there exists a unique martingale measure $\mathbb{Q}^{XM,d}$ such that all $X^{(d/f)}M^{(f)}$ rebased traded assets in the domestic economy become martingales. What is the relation between the probability measure $\mathbb{Q}^{XM,d}$ in the domestic economy and $\mathbb{Q}^{M,f}$ in the foreign economy? All $X^{(d/f)}M^{(f)}$ rebased traded assets in the domestic economy are martingales under $\mathbb{Q}^{XM,d}$. Hence, also the domestic value of the foreign traded assets: $X^{(d/f)}Z^{(f)}/X^{(d/f)}M^{(f)} = Z^{(f)}/M^{(f)}$ are martingales. But this implies (given the uniqueness of a probability measure for a particular numeraire) that $\mathbb{Q}^{XM,d}$ and $\mathbb{Q}^{M,f}$ are the same probability measure. So under the measure $\mathbb{Q}^{M,f}$ all $X^{(d/f)}M^{(f)}$ rebased traded assets are martingales in the domestic economy and all $M^{(f)}$ rebased traded assets are martingales in the foreign economy.

From the domestic economy perspective we have $N^{(d)}$ and $X^{(d/f)}M^{(f)}$ as domestic numeraires. Hence, we can apply the single currency Change of Numeraire Theorem which yields

$$\frac{d\mathbb{Q}^{N,d}}{d\mathbb{Q}^{M,f}} = \frac{N^{(d)}(T)}{X^{(d/f)}(T)M^{(f)}(T)}\frac{X^{(d/f)}(0)M^{(f)}(0)}{N^{(d)}(0)}. \tag{11.2}$$

On the other hand, from the foreign perspective we have $X^{(f/d)}N^{(d)}$ and $M^{(f)}$ as foreign numeraires. Hence, we can apply the single currency Change of Numeraire Theorem to this case as well which yields

$$\frac{d\mathbb{Q}^{N,d}}{d\mathbb{Q}^{M,f}} = \frac{X^{(f/d)}(T)N^{(d)}(T)}{M^{(f)}(T)} \frac{M^{(f)}(0)}{X^{(f/d)}(0)N^{(d)}(0)}. \tag{11.3}$$

Note that equation (11.2) and (11.3) are identical since $X^{(f/d)} = 1/X^{(d/f)}$. Hence, we obtain:

Theorem (Multi-Currency Change of Numeraire) *Given an arbitrage-free system of economies (d, f), an exchange rate X and two numeraires $N^{(d)}$ and $M^{(f)}$ within the economies with the associated martingale measures $\mathbb{Q}^{N,d}$ and $\mathbb{Q}^{M,f}$ we have that*

$$\begin{aligned}
\frac{d\mathbb{Q}^{N,d}}{d\mathbb{Q}^{M,f}} &= \frac{N^{(d)}(T)}{X^{(d/f)}(T)M^{(f)}(T)} \frac{X^{(d/f)}(0)M^{(f)}(0)}{N^{(d)}(0)} \\
&= \frac{X^{(f/d)}(T)N^{(d)}(T)}{M^{(f)}(T)} \frac{M^{(f)}(0)}{X^{(f/d)}(0)N^{(d)}(0)}.
\end{aligned} \tag{11.4}$$

Proof. A more formal proof of this theorem can be found in Musiela and Rutkowski (1997, Chapter 17). □

Note, that the Single Currency Change of Numeraire Theorem can be seen as a special case of the Multi-Currency Theorem with $X \equiv 1$ (and $f = d$). Musiela and Rutkowski (1997, Chapter 17) also discuss the multi-currency generalisation of the market models of Chapter 8.

11.1.2 Convexity Correction

Given the Change of Numeraire Theorems, we are now in a position to investigate convexity correction. Convexity correction arises when an interest rate is paid out at the "wrong" time and/or in the "wrong" currency.

Single Currency. Suppose we are given a forward interest rate y with maturity T (which can be, for example, a forward LIBOR rate or a forward par swap rate) and a numeraire P such that the forward rate y is a martingale under the associated probability measure $\mathbb{Q}^P$. In the single currency case we can have a contract where the interest rate $y(T)$ is observed at T but paid at a later date $S \geq T$. If we denote the discount bond that matures at S by D_S, we have that the value of this contract at time T is given by $V(T) = y(T)D_S(T)$. Using D_S as the numeraire with the associated forward measure $\mathbb{Q}^S$ we can express the value of this contract at time 0 as

$$V(0) = D_S(0)\mathbb{E}^S\left(y(T)\right). \tag{11.5}$$

However, under the measure $\mathbb{Q}^S$ the process y is in general not a martingale. Expectation (11.5) can be expressed as $y(0)$ times a correction term. This correction term is known in the market as the convexity correction. Using (11.5) market participants calculate the value of the payoff as the discounted value of the convexity corrected payoff $\mathbb{E}^S(y)$. What remains to be done, is to find an expression for $\mathbb{E}^S(y)$.

We do know the process of y under its "own" martingale measure $\mathbb{Q}^P$. Using the Change of Numeraire Theorem, we can express the expectation $\mathbb{E}^S$ in terms of $\mathbb{E}^P$ as follows

$$\mathbb{E}^S\left(y(T)\right) = \mathbb{E}^P\left(y(T)\frac{d\mathbb{Q}^S}{d\mathbb{Q}^P}\right) = \mathbb{E}^P\left(y(T)\frac{D_S(T)}{P(T)}\frac{P(0)}{D_S(0)}\right) = \mathbb{E}^P\left(y(T)R(T)\right)$$
(11.6)

where R denotes the Radon-Nikodym derivative. Both y and R are martingales under the measure $\mathbb{Q}^P$. If we know the joint law of y and R the expectation can be calculated explicitly and we obtain an expression for the convexity correction.

One possible approach is to assume that both y and R are lognormal. Hence, we assume that y follows the process $dy = \sigma_y y\, dW_y$ and R follows the process $dR = \sigma_R R\, dW_R$ under the measure $\mathbb{Q}^P$ with correlation $dW_y\, dW_R = \rho_{y,R}\, dt$. Under this assumption we can calculate (11.6) as

$$\mathbb{E}^S\left(y(T)\right) = \mathbb{E}^P\left(y(T)R(T)\right) = y(0)e^{\rho_{y,R}\sigma_y\sigma_R T}.$$
(11.7)

Note that the Radon-Nikodym derivative R is a ratio of traded assets whose values can be observed in the market. Hence, the instantaneous volatility and the correlation of the R process (which remain unaffected by a change of measure) can be estimated from historical data.

Another approach is to approximate R by a function of a random variable whose behaviour is known, for example the rate y itself. To check the accuracy of the convexity correction, one can compare the values of the approximating model with the values obtained by (11.7).

If the numeraire P is a PVBP of the form $\sum_i \alpha_{i-1} D_i$ (which is the case for interest rate derivatives we consider in this chapter) we have a convenient way to approximate $R(T)$ via the Linear Swap Rate Model (see the Appendix at the end of this chapter for a derivation of this model). In this model one approximates $D_S(T)/P(T)$ by the linear form $A + B_S y(T)$, where $A = (\sum_i \alpha_{i-1})^{-1}$ and $B_S = (D_S(0)/P(0) - A)/y(0)$. Using this approximation we can evaluate (11.6) explicitly as

$$\begin{aligned}
\mathbb{E}^S\left(y(T)\right) &= \frac{P(0)}{D_S(0)}\mathbb{E}^P\left(y(T)\frac{D_S(T)}{P(T)}\right) \\
&= \frac{P(0)}{D_S(0)}\mathbb{E}^P\left(y(T)\left(A + B_S y(T)\right)\right) \\
&= y(0)\left(\frac{A + B_S y(0)e^{\sigma_y^2 T}}{A + B_S y(0)}\right).
\end{aligned}$$
(11.8)

If we equate the convexity correction expressions (11.7) and (11.8), we obtain

$$\rho_{y,R}\sigma_y\sigma_R T = \log\left(\frac{A + B_S y(0)e^{\sigma_y^2 T}}{A + B_S y(0)}\right). \tag{11.9}$$

This expression can be tested empirically using historical data, and provides a method to assess the accuracy of the Linear Swap Rate Model for calculating convexity corrections.

Cross currency. Let us now consider the cross currency case. In this case we can have a contract where a foreign interest rate $y^{(f)}(T)$ is observed at T but paid in domestic currency at a later date $S \geq T$. If we denote the domestic discount bond that matures at S by $D_S^{(d)}$, we have that the value of this contract in domestic terms at time T is given by $V^{(d)}(T) = y^{(f)}(T)D_S^{(d)}(T)$. Using $D_S^{(d)}$ as the numeraire and the associated the measure $\mathbb{Q}^{S,d}$ we can express the value of this contract at time 0 as

$$V^{(d)}(0) = D_S^{(d)}(0)\mathbb{E}^{S,d}\left(y^{(f)}(T)\right). \tag{11.10}$$

However, under the measure $\mathbb{Q}^{S,d}$ the process $y^{(f)}$ is in general not a martingale. We do know the process of $y^{(f)}$ under its "own" martingale measure $\mathbb{Q}^{P,f}$. Using the Change of Numeraire Theorem, we can express the expectation $\mathbb{E}^{S,d}$ in terms of $\mathbb{E}^{P,f}$ as follows

$$\mathbb{E}^{S,d}\left(y^{(f)}(T)\right) = \mathbb{E}^{P,f}\left(y^{(f)}(T)\frac{d\mathbb{Q}^{S,d}}{d\mathbb{Q}^{P,f}}\right) = \mathbb{E}^{P,f}\left(y^{(f)}(T)R^{(d/f)}(T)\right). \tag{11.11}$$

One possible approach to evaluate (11.11) is to assume that both $y^{(f)}$ and $R^{(d/f)}$ are lognormal martingales with (constant) volatilities σ_y and σ_R respectively. Under this assumption we obtain

$$\mathbb{E}^{S,d}\left(y^{(f)}(T)\right) = \mathbb{E}^{P,f}\left(y^{(f)}(T)R^{(d/f)}(T)\right) = y(0)e^{\rho_{y,R}\sigma_y\sigma_R T}, \tag{11.12}$$

where $\rho_{y,R}$ denotes the correlation between $y^{(f)}$ and $R^{(d/f)}$. Just like in the single currency case the Radon-Nikodym derivative $R^{(d/f)}$ is a ratio of traded assets whose values can be observed in the market. Hence, the volatility and the correlation of the $R^{(d/f)}$ process (which remain unaffected by a change of measure) can be estimated from historical data.

Another approach to evaluate (11.11) is to decompose $R^{(d/f)}$ as the forward exchange rate times $D_S^{(f)}/P^{(f)}$, which is the single currency Radon-Nikodym derivative of equation (11.6). This leads to the expression

$$\mathbb{E}^{S,d}\left(y^{(f)}(T)\right)$$
$$= \mathbb{E}^{P,f}\left(y^{(f)}(T)\frac{d\mathbb{Q}^{S,d}}{d\mathbb{Q}^{P,f}}\right)$$

$$= \mathbb{E}^{P,f}\left(y^{(f)}(T)\left(X^{(f/d)}(T)\frac{D_S^{(d)}(T)}{P^{(f)}(T)}\right)\right)\frac{P^{(f)}(0)}{X^{(f/d)}(0)D_S^{(d)}(0)}$$

$$= \mathbb{E}^{P,f}\left(y^{(f)}(T)\left(X^{(f/d)}(T)\frac{D_S^{(d)}(T)}{D_S^{(f)}(T)}\right)\frac{D_S^{(f)}(T)}{P^{(f)}(T)}\right)\frac{P^{(f)}(0)}{X^{(f/d)}(0)D_S^{(d)}(0)}$$

$$= \mathbb{E}^{P,f}\left(y^{(f)}(T)F_S^{(f/d)}(T)\frac{D_S^{(f)}(T)}{P^{(f)}(T)}\right)\frac{P^{(f)}(0)}{X^{(f/d)}(0)D_S^{(d)}(0)}, \tag{11.13}$$

where F_S denotes the forward exchange rate for delivery at time S. The volatility of the forward exchange rate is quoted in the market as the implied volatility of a foreign exchange option with a maturity of S years. The process F_S is not a martingale under the measure $\mathbb{Q}^{P,f}$, but the drift of the process can be determined from the fact that $D_S^{(f)}/P^{(f)}$ and $F_S D_S^{(f)}/P^{(f)}$ are martingales under $\mathbb{Q}^{P,f}$.

Let us consider once more using the Linear Swap Rate Model. In the foreign economy, we approximate $D_S^{(f)}/P^{(f)}$, by $A^{(f)} + B_S^{(f)}y^{(f)}$. First, we determine the expected value of F from the fact that the Radon-Nikodym derivative $R^{(d/f)}$ is a martingale with expected value 1. This leads to

$$\mathbb{E}^{P,f}\left(F_S^{(f/d)}(T)\right) = F_S^{(f/d)}(0)\left(\frac{A^{(f)} + B_S^{(f)}y^{(f)}(0)}{A^{(f)} + B_S^{(f)}y^{(f)}(0)e^{\rho_{F,y}\sigma_F\sigma_y T}}\right) \tag{11.14}$$

where we have made the (market standard) assumption that F is a lognormal process, furthermore the F volatility is denoted by σ_F and $\rho_{F,y}$ denotes the correlation between F and y. We can now evaluate the cross currency expectation (11.13) as

$$\mathbb{E}^{S,d}\left(y^{(f)}(T)\right) = y^{(f)}(0)\left(e^{\rho_{F,y}\sigma_F\sigma_y T}\frac{A^{(f)} + B_S^{(f)}y^{(f)}(0)e^{(\sigma_y^2 + \rho_{F,y}\sigma_F\sigma_y)T}}{A^{(f)} + B_S^{(f)}y^{(f)}(0)e^{\rho_{F,y}\sigma_F\sigma_y T}}\right). \tag{11.15}$$

Note that this expression depends on the F and the $y^{(f)}$ volatility which can both be observed in the market, also the correlation $\rho_{F,y}$ between the forward F/X process and the forward swap rate process enters the formula.

Note also that, if we set $\rho_{F,y} = 0$ or $\sigma_F = 0$, then the cross currency formula (11.15) reduces to the single currency formula (11.8).

11.2 Options on Convexity Corrected Rates

Not only are we interested in interest rates paid in different currencies and/or at different times, but also we want to value call and put options and digital options on these rates. Using the Change of Numeraire Theorem, we can proceed exactly as before.

11.2.1 Option Price Formula

Take for example a call option on a foreign interest rate observed at time T and paid in domestic currency at a later date $S \geq T$. The value of such a contract at time T is given by $V^{(d)}(T) = D_S^{(d)}(T)\max\{y^{(f)}(T) - K, 0\}$. We can value this payoff as

$$V^{(d)}(0) = D_S^{(d)}(0)\mathbb{E}^{S,d}\left(\max\{y^{(f)}(T) - K, 0\}\right) \tag{11.16}$$

$$= D_S^{(d)}(0)\mathbb{E}^{P,f}\left(\max\{y^{(f)}(T) - K, 0\}R^{(d/f)}(T)\right) \tag{11.17}$$

$$= D_S^{(d)}(0)\mathbb{E}^{P,f}\left(\max\{y^{(f)}(T)R^{(d/f)}(T) - KR^{(d/f)}(T), 0\}\right). \tag{11.18}$$

The last line can be interpreted as an exchange option between yR and KR. We know that y is a lognormal martingale under the measure $\mathbb{Q}^{P,f}$. Under the assumption that R has a lognormal distribution as well then also yR will have a lognormal distribution. The option value can then be evaluated by the Margrabe (1978) formula as follows:

$$V^{(d)}(0) = D_S^{(d)}(0)\left(E_1 N(d_1) - E_2 N(d_2)\right),$$
$$E_1 = \mathbb{E}^{P,f}\left(y^{(f)}(T)R^{(d/f)}(T)\right),$$
$$E_2 = \mathbb{E}^{P,f}\left(KR^{(d/f)}(T)\right) = K,$$
$$d_{1,2} = \frac{\log(E_1/E_2) \pm \frac{1}{2}\Sigma^2}{\Sigma}, \tag{11.19}$$
$$\Sigma^2 = \mathrm{Var}^{P,f}\left(\log\left(\frac{y^{(f)}(T)R^{(d/f)}(T)}{KR^{(d/f)}(T)}\right)\right)$$
$$= \mathrm{Var}^{P,f}\left(\log y^{(f)}(T)\right) = \sigma_y^2 T.$$

The expression for E_1 is the convexity corrected forward rate which can be denoted by $\tilde{y}^{(f)}$. Hence, we can simplify the expression above to obtain

$$V^{(d)}(0) = D_S^{(d)}(0)\left(\tilde{y}^{(f)}N(d_1) - KN(d_2)\right),$$
$$d_{1,2} = \frac{\log(\tilde{y}^{(f)}/K) \pm \frac{1}{2}\sigma_y^2 T}{\sigma_y\sqrt{T}}, \tag{11.20}$$

which is the market standard method of valuing options on convexity corrected rates: apply the Black formula using the convexity corrected rate as the forward rate.

We could have obtained formula (11.20) directly by evaluating (11.16). This shows (in an indirect way) that the probability distribution of $y^{(f)}$ is also a lognormal distribution under $\mathbb{Q}^{S,d}$ with the same volatility σ_y but with a mean of $\tilde{y}^{(f)}$.

For the cases where we approximate the Radon-Nikodym derivative with the Linear Swap Rate Model, the approximation results in an expression for the Radon-Nikodym derivative which is not exactly lognormally distributed.

However, the market standard method can still be used as a good approximation.

11.2.2 Digital Price Formula

A digital call option on a foreign interest rate observed at time T and paid in domestic currency at a later date $S \geq T$. has a value at time T of $V^{\mathrm{DCALL},(d)}(T) = D_S^{(d)}(T)\mathbb{1}\big(y^{(f)}(T) > K\big)$, where $\mathbb{1}()$ denotes the indicator function.

We can calculate the value of such a digital call as

$$V^{\mathrm{DCALL},(d)}(0) = D_S^{(d)}(0)N(d_2) \tag{11.21}$$

and the value of a digital put option as

$$V^{\mathrm{DPUT},(d)}(0) = D_S^{(d)}(0)N(-d_2), \tag{11.22}$$

where the $\tilde{d}$ are defined in (11.20).

11.3 Single Index Products

Using the theoretical results derived in Sections 11.1 and 11.2, we will derive pricing formulæ for a variety of products whose payoff is determined by a single interest rate.

11.3.1 LIBOR in Arrears

In a normal LIBOR payment, the LIBOR interest rate L_i is observed at time T_i and paid at the end of the accrual period at time T_{i+1} as $\alpha_i L_i(T_i)$, where α_i denotes the daycount fraction. The forward LIBOR rate $L_i(t)$ is defined as $(D_i(t) - D_{i+1}(t))/\alpha_i D_{i+1}(t)$, where D_i denotes the discount factor that matures at time T_i. Hence, if we choose D_{i+1} as the numeraire then under the associated martingale measure $\mathbb{Q}^{i+1}$ the forward LIBOR rate L_i is a martingale.

In a LIBOR *in Arrears* contract, the interest payment at time T_{i+1} is based on $L_{i+1}(T_{i+1})$. Hence, the LIBOR contract is fixed only at the end of the interest rate period, i.e. is fixed in arrears.

Payment. The payoff of a LIBOR in Arrears payment at time T_i is therefore equal to $V^{\mathrm{LIA}}(T_i) = L_i(T_i)$. The value at time 0 of this payment is given by $V^{\mathrm{LIA}}(0) = D_i(0)\mathbb{E}^i\big(L_i(T_i)\big) = D_i(0)\tilde{L}_i$, where $\tilde{L}_i$ denotes the convexity correction on the LIBOR rate to reflect the payment in arrears. To calculate the convexity correction we use formula (11.8). Note, that in the case of

Table 11.1. Pricing errors (in bp) for LIBOR in Arrears.

	$D_T(0) = e^{-0.10T}; K = 10\%$			$D_T(0) = e^{-0.05T}; K = 5\%$		
Mat	$\sigma = 15\%$	$\sigma = 10\%$	$\sigma = 5\%$	$\sigma = 15\%$	$\sigma = 10\%$	$\sigma = 5\%$
1D	0.00	0.00	0.00	0.00	0.00	0.00
1M	0.00	0.00	0.00	0.00	0.00	0.00
3M	0.00	0.00	0.00	0.00	0.00	0.00
6M	0.00	0.00	0.00	0.00	0.00	0.00
9M	0.00	0.00	0.00	0.00	0.00	0.00
1Y	0.00	0.00	0.00	0.00	0.00	0.00
2Y	0.00	0.00	0.00	0.00	0.00	0.00
3Y	0.02	0.00	0.00	0.00	0.00	0.00
4Y	0.03	0.00	0.00	0.02	0.00	0.00
5Y	0.04	0.00	0.00	0.02	0.00	0.00
6Y	0.06	0.01	0.00	0.03	0.00	0.00
7Y	0.08	0.02	0.00	0.04	0.00	0.00
8Y	0.10	0.02	0.00	0.05	0.00	0.00
9Y	0.12	0.02	0.00	0.06	0.00	0.00
10Y	0.16	0.02	0.00	0.08	0.00	0.00

LIBOR in Arrears the Linear Swap Rate Model is exact with $y = L_i$, $A = 1$, $B_{T_i} = \alpha_i$ and we obtain

$$\tilde{L}_i = L_i(0) \left(\frac{1 + \alpha_i L_i(0) e^{\sigma_i^2 T_i}}{1 + \alpha_i L_i(0)} \right) \tag{11.23}$$

where σ_i denotes the volatility of L_i.[17]

Option. A LIBOR in Arrears caplet has a payoff at time T_i of $V^{\text{LIACAP}}(T_i) = \max\{L_i(T_i) - K, 0\}$. Using the set up of Section 11.2, we can approximate the value of a LIBOR in Arrears caplet at time 0 with formula (11.20).

Because the Linear Swap Rate Model is exact in the case of LIBOR in Arrears, we know in this special case the exact pricing formula V^{XLIACAP}. For a LIBOR in Arrears caplet this formula is given by the expectation (11.16) which can be evaluated explicitly as

$$V^{\text{XLIACAP}}(0) = D_i(0) \big(\omega B(L_i(0), \sigma_i, K, T_i) + (1 - \omega) B(L_i^*, \sigma_i, K, T_i) \big),$$
$$L_i^* = L_i(0) e^{\sigma_i^2 T_i},$$
$$\omega = \frac{D_{i+1}(0)}{D_i(0)},$$

$$\tag{11.24}$$

where $B()$ denotes the forward Black formula. The exact pricing formula gives us the opportunity to check the accuracy of the approximate pricing formula (11.20).

---

[17]Note that the LIBOR in Arrears convexity correction formula given in Hull (2000, p. 553) is to first order equal to the exact formula (11.23).

In Table 11.1 we have compared for different maturities the pricing differences of $V^{\text{XLIACAP}} - V^{\text{LIACAP}}$ in basis points. We have used the following inputs: annual LIBOR with $\alpha \equiv 1$, initial yield-curves flat at 5% and 10% and a flat volatility structures.

From the table we see that the approximating pricing formula is very accurate. For almost all entries in the table the error is less than 0.1bp.

11.3.2 Constant Maturity Swap

A *Constant Maturity Swap* (or CMS) payment consists of a swap rate y that is observed at time T and paid out (only once) at time $S \geq T$. The forward swap rate $y(t)$ is defined as $(D_0(t) - D_N(t))/P(t)$, where D_0 has a maturity equal to the start date T of the swap, D_N has a maturity equal to the last payment date T_N of the swap and $P(t)$ is the PVBP of the swap given by $\sum_i \alpha_{i-1} D_i(t)$ where the D_i are the discount factors with maturity dates equal to the dates at which fixed payments are made in the swap. If we use the PVBP $P(t)$ as a numeraire, then the forward swap rate $y(t)$ is a martingale under the measure $\mathbb{Q}^P$.

Convexity Correction. To calculate the value of a CMS payment at time 0 we use the convexity corrected swap rate $\tilde{y}$ given by (11.8).

Let us compare the convexity correction formula (11.8) with the correction formula (20.19) given in Hull (2000, p. 554). To do this we consider Example 20.8 of Hull (2000, p. 555). In this example, a 5-year semi-annual swap rate is observed in 4 years time and paid 6 months later. The term-structure of interest rates is flat at 5% per annum with semi-annual compounding and $\sigma_y = 15\%$. The daycount fractions are assumed to be equal to 0.5. To calculate the convexity correction we compute the following numbers:

$$D_T = 1.025^{-2T}$$

$$P = \sum_{i=1}^{10} 0.5 D_{4+i/2} = 3.59161$$

$$y = \frac{D_4 - D_9}{P} = 0.05$$

$$A = \frac{1}{\sum_{i=1}^{10} 0.5} = 0.2 \qquad (11.25)$$

$$B = \frac{D_{4.5}/P - A}{y} = 0.458879$$

$$\tilde{y} = y \left(\frac{A + Bye^{0.15^2 4}}{A + By} \right) = 0.05048$$

The convexity correction of +4.8bp we find with the Linear Swap Rate Model is very close to the correction of +4.5bp reported by Hull.

11.3.3 Diffed LIBOR

For a *diffed* LIBOR contract (also known as quantoed LIBOR) a foreign LIBOR rate $L^{(f)}$ is observed at T_i and is paid in domestic currency at time S, with $T_i \leq S \leq T_{i+1}$. For $S = T$, the contract reduces to diffed LIBOR in Arrears and for $S = T_{i+1}$, the contract reduces to a diffed standard LIBOR payment. These diffed LIBOR payments are often paid in the form of a *differential swap* or diff swap. In a diff swap a floating interest rate in the foreign currency is exchanged against a floating rate in the domestic currency where both rates are applied to the same domestic notional principal. The domestic leg of a diff swap can be valued as a standard floating leg in the domestic currency. The foreign leg of the diff swap is a portfolio of diffed foreign LIBOR payments, and its value can be calculated as the sum of the individual diffed LIBOR payments.

To obtain a valuation formula for the diffed LIBOR contract, we consider the special case of the Linear Swap Rate Model of Section 11.3.2 for the foreign economy.

Convexity Correction. To calculate the value of a diffed LIBOR payment at time 0 we use the convexity corrected foreign LIBOR rate $\tilde{L}_i^{(f)}$ given by

$$
\tilde{L}_i^{(f)} = L_i^{(f)}(0) \left(e^{\rho_{F,i}\sigma_F\sigma_i T_i} \frac{1 + B_S^{(f)} L_i^{(f)}(0) e^{(\sigma_i^2 + \rho_{F,i}\sigma_F\sigma_i)T_i}}{1 + B_S^{(f)} L_i^{(f)}(0) e^{\rho_{F,i}\sigma_F\sigma_i T_i}} \right), \qquad (11.26)
$$

with $B_S^{(f)}$ given in by the Linear Swap Rate Model. Furthermore the volatility of the forward exchange rate with delivery at time S is denoted by σ_F, σ_i denotes the volatility of $L^{(f)}$ and $\rho_{F,i}$ denotes the correlation between the forward (f/d) exchange rate and $L^{(f)}$. By the exchange rate (f/d) we mean the value of 1 unit of domestic currency in terms of foreign currency. For many practical applications, the correlation between the forward exchange rate and the forward LIBOR rate is approximated by the correlation between the spot exchange rate and the spot LIBOR rate. Note that, if we calculate the correlation between the forward (d/f) exchange rate and $L^{(f)}$, we must use $-\rho$ in our formulæ.

11.3.4 Diffed CMS

For a *diffed CMS* contract (also known as quantoed CMS) a foreign swap rate $y^{(f)}$ is observed at T and is paid in domestic currency at time S, with $T \leq S$.

Convexity Correction. To calculate the value of a diffed CMS payment at time 0 we use the convexity corrected swap rate given by (11.15).

11.4 Multi-Index Products

In this section we will give an overview of interest rate derivatives whose payoff depends on multiple underlying values. Although the payoff for the products is more complicated, we can still use the framework developed in Sections 11.1 and 11.2.

11.4.1 Rate Based Spread Options

A *rate based spread option* pays out a weighted sum of two interest rates, if this sum exceeds a certain strike K, where K can be positive, zero or negative. The observed interest rates can either be in the domestic or in (possibly two different) foreign economies and can be observed at different points in time U and T, and the payment is received at time S. The payoff function $V^{\mathrm{SPR},d}$ in domestic terms is given by:

$$V^{\mathrm{SPR},d}(T) = D_S^{(d)}(T) \max\{ay^{(1)}(U) + by^{(2)}(T) - K, 0\} \qquad (11.27)$$

where $U \le T \le S$, a and b are arbitrary weighting factors, K is the strike, $D_S^{(d)}(T)$ is the value at time T of a (domestic) discount factor with maturity S and $y^{(1)}$ and $y^{(2)}$ are two swap or LIBOR rates observed in domestic or foreign economies. Hence, the rates are observed at $U \le T$ and the payment is made at time $S \ge T$. Note, that one can generate a PVBP-based payoff (for example in a swap where the fixed side is determined at the start date by a spread between interest rates) by adding several terms with the same observation dates U, T but with different payment dates S. An example are *quanto swaptions* with payoff $P^{(d)}(T) \max\{y^{(d)}(T) - y^{(f)}(T), 0\}$, where $P^{(d)}$ denotes a PVBP in domestic currency. Spread options of this form have been analysed by Hunt and Pelsser (1998).

The option value can be computed using the framework outlined in Sections 11.1 and 11.2 as

$$V^{\mathrm{SPR},d}(0) = D_S^{(d)}(0)\mathbb{E}^{S,d}\left(\max\{ay^{(1)}(U) + by^{(2)}(T) - K, 0\}\right). \qquad (11.28)$$

The processes for $y^{(1)}$ and $y^{(2)}$ are no longer martingales under the measure $\mathbb{Q}^{S,d}$. Note that we have to be a bit careful here since $U \le T$. The option payoff is determined by the joint probability distribution of the random variables $y^{(1)}(U)$ and $y^{(2)}(T)$. Using the expressions derived in Sections 11.1 and 11.2 we can obtain expressions for the means of $y^{(1)}$ and $y^{(2)}$ under the measure $\mathbb{Q}^{S,d}$. Note, that we also ought to calculate the correlation of $y^{(1)}(U)$ and $y^{(2)}(T)$ under the measure $\mathbb{Q}^{S,d}$. The expression we would calculate would consist of the instantaneous correlation $\rho_{1,2}$ with a convexity correction term. The instantaneous correlation has to be obtained from historical estimation. However, the historical estimation of correlations is not a very accurate procedure. We will therefore ignore the convexity correction for the correlation and use the historical estimate directly for our calculations.

If we now make the assumption that under the measure $\mathbb{Q}^{S,d}$ both $y^{(1)}$ and $y^{(2)}$ have lognormal distributions (with expected values $\tilde{y}^{(1)}$, $\tilde{y}^{(2)}$, volatilities $\sigma_{(1)}$, $\sigma_{(2)}$ and correlation $\rho_{1,2}$), then the option value can be computed by representing the expectation $\mathbb{E}^{S,d}$ in (11.28) as

$$\int_{-\infty}^{\infty} \int_{-\infty}^{\infty} \max\{\alpha e^{\sigma_1 \epsilon_1} + \beta e^{\sigma_2(\rho \epsilon_1 + \sqrt{1-\rho^2}\epsilon_2)} - K, 0\} \frac{e^{-\frac{1}{2}\epsilon_2^2}}{\sqrt{2\pi}} \frac{e^{-\frac{1}{2}\epsilon_1^2}}{\sqrt{2\pi}} \, d\epsilon_2 \, d\epsilon_1,$$

(11.29)

where

$$\alpha = a\tilde{y}^{(1)} e^{-\frac{1}{2}\sigma_1^2}$$

$$\beta = b\tilde{y}^{(2)} e^{-\frac{1}{2}\sigma_2^2}$$

$$\sigma_1 = \sigma_{(1)}\sqrt{U}$$

(11.30)

$$\sigma_2 = \sigma_{(2)}\sqrt{T}$$

$$\rho = \rho_{1,2}\sqrt{\frac{U}{T}}.$$

The double integral can be calculated more efficiently by first solving the inner integral analytically and then numerically integrating the remaining outer integral. The inner integral can be solved as follows. First we have to determine when the expression inside the max operator is positive, hence we must solve $\alpha e^{\sigma_1 \epsilon_1} + \beta e^{\sigma_2(\rho \epsilon_1 + \sqrt{1-\rho^2}\epsilon_2)} - K > 0$. This is an inequality in ϵ_1 and ϵ_2. For $\beta > 0$ the inequality can be solved as

$$\epsilon_2 > \frac{\log^*\left(\frac{K - \alpha e^{\sigma_1 \epsilon_1}}{\beta e^{\rho \sigma_2 \epsilon_1}}\right)}{\sigma_2\sqrt{1-\rho^2}},$$

(11.31)

where $\log^*()$ is equal to the logarithm function if the argument is positive and equal to $-\infty$ if the argument is negative. To lighten the notation we denote the inequality as $\epsilon_2 > l^*(\epsilon_1)$. We can now calculate the inner integral of (11.29) analytically and we obtain the following single integral

$$\int_{-\infty}^{\infty} (\alpha e^{\sigma_1 \epsilon_1} - K) N\left(-l^*(\epsilon_1)\right) \frac{e^{-\frac{1}{2}\epsilon_1^2}}{\sqrt{2\pi}} \, d\epsilon_1$$

$$+ \beta e^{\frac{1}{2}\sigma_2^2(1-\rho^2)} \int_{-\infty}^{\infty} e^{\rho \sigma_2 \epsilon_1} N\left(-l^*(\epsilon_1) + \sigma_2\sqrt{1-\rho^2}\right) \frac{e^{-\frac{1}{2}\epsilon_1^2}}{\sqrt{2\pi}} \, d\epsilon_1 \quad (11.32)$$

that can be evaluated via numerical integration.

For $\beta < 0$ the inequality $\alpha e^{\sigma_1 \epsilon_1} + \beta e^{\sigma_2(\rho \epsilon_1 + \sqrt{1-\rho^2}\epsilon_2)} - K > 0$ can be solved as

$$\epsilon_2 < \frac{\log^*\left(\frac{\alpha e^{\sigma_1 \epsilon_1} - K}{-\beta e^{\rho \sigma_2 \epsilon_1}}\right)}{\sigma_2\sqrt{1-\rho^2}}.$$

(11.33)

Denoting the inequality as $\epsilon_2 < l^*(\epsilon_1)$ we can express (11.29) as

$$\int_{-\infty}^{\infty} (\alpha e^{\sigma_1 \epsilon_1} - K) N\big(l^*(\epsilon_1)\big) \frac{e^{-\frac{1}{2}\epsilon_1^2}}{\sqrt{2\pi}} \, d\epsilon_1$$

$$+ \beta e^{\frac{1}{2}\sigma_2^2(1-\rho^2)} \int_{-\infty}^{\infty} e^{\rho\sigma_2\epsilon_1} N\big(l^*(\epsilon_1) - \sigma_2\sqrt{1-\rho^2}\big) \frac{e^{-\frac{1}{2}\epsilon_1^2}}{\sqrt{2\pi}} \, d\epsilon_1. \quad (11.34)$$

Note that for $K = 0$ the function $l^*(\epsilon_1)$ reduces to a linear function in ϵ_1. For this special case the integrals given above can be solved analytically and yield generalisations of the Margrabe (1978) formula.

11.4.2 Spread Digital

A rate based spread digital option has the following payoff function:

$$V^{\text{SPRDIG},d}(T) = D_S^{(d)}(T)\mathbb{1}\big(a y^{(1)}(U) + b y^{(2)}(T) - K > 0\big) \qquad (11.35)$$

where $\mathbb{1}()$ denotes the indicator function. Note that this expression is the derivative with respect to $-K$ of (11.27). Hence, we can calculate the digital option value at time 0 as (using the same notation as before)

$$D_S^{(d)}(0) \int_{-\infty}^{\infty} N\big(-l^*(\epsilon_1)\big) \frac{e^{-\frac{1}{2}\epsilon_1^2}}{\sqrt{2\pi}} \, d\epsilon_1 \qquad (11.36)$$

for $b > 0$ or as

$$D_S^{(d)}(0) \int_{-\infty}^{\infty} N\big(l^*(\epsilon_1)\big) \frac{e^{-\frac{1}{2}\epsilon_1^2}}{\sqrt{2\pi}} \, d\epsilon_1 \qquad (11.37)$$

for $b < 0$.

11.4.3 Other Multi-Index Products

The spread options and spread digital options discussed here are only a few examples of possible exotic interest rate derivatives. Other example are *trigger swaptions* which can be viewed as outside barrier swaptions, where the swaption payoff $\max\{y^{(1)} - K, 0\}$ is only received if another rate $y^{(2)}$ is above or below a certain level H. A variant on these trigger products are *floating accrual* notes where a LIBOR rate $L_i(T_i)$ is observed at the beginning of the accrual period, but the payment at the end of the period T_{i+1} is based on the number of days that the spot LIBOR rate for that day fixed above or below a certain level H. This kind of product can be decomposed as a portfolio of trigger caplets with payoff $D_{i+1}(S)L_i(T_i)\mathbb{1}(L_S(S) > H)$ for $T_i \leq S \leq T_{i+1}$.

All these type of products can be analysed using the framework outlined here.

11.4.4 Comparison with Market Models

The spread options which we a considering here can, of course, also be valued using the market models of Chapter 8 or 9. Let us consider the case of a LIBOR spread option. A single currency LIBOR spread option has as payoff at time T_i of $V(T_i) = \alpha_i \max\{L_i(T_i) - L_{i-1}(T_{i-1}), 0\}$ where all variables are denominated in the same currency. This product can be considered to be a special case of a ratchet option, which was treated in Chapter 8, Section 8.4.3. On the other hand, LIBOR spread options can be valued as a single currency spread option.

Using curves and volatilities on November 15, 1999 we priced a spread option on 6M LIBOR in EUR. We calculated the value of this LIBOR spread option that uses the 6M LIBOR fixings at November 15, 2004 and May 13, 2005, which pays out at May 17, 2005. Since there is no strike in this example, we can price this option as an exchange option using the Margrabe (1978) formula.

Using the LIBOR market model, it may seem we are pricing a two-factor payoff (the spread option) with a one-factor model. However, this is where the mean-reversion parameter comes in. As we are observing the two LIBOR rates at two different points in time, we can use the mean-reversion parameter to control the correlation between the two LIBOR rates.

For different mean-reversion levels (and the corresponding correlations implied by those mean-reversion levels), we have reported the values of spread caps and spread floors in Table 11.2.

Table 11.2. Prices (in bp) of LIBOR spread options.

MeanRev	-10%		0%		10%	
Corr	0.974		0.954		0.927	
	LMM	SPR	LMM	SPR	LMM	SPR
SprCap	13.2	13.2	15.8	15.8	18.7	18.7
SprFlr	5.1	5.1	7.8	7.7	10.6	10.6

We see from the table that we find a very close correspondence between the MC simulation in the LIBOR market model (LMM) and the spread option formula (SPR). The advantage of using the spread option approach of this chapter is that it is much faster and more accurate than the Monte Carlo implementation of the LIBOR market model. Also, using the spread option approach we focus much more on the relevant LIBOR rates and the relevant correlation between the two rates under consideration. Using the LIBOR market model, the correlation between the two relevant LIBOR rates is controlled indirectly via the mean-reversion parameter.

11.5 A Warning on Convexity Correction

For all the convexity corrections we have calculated we have made the assumption that the rates under consideration have a lognormal probability distribution. Using the lognormal distribution is, of course, the market standard. However, the magnitude of the convexity correction calculated does depend on the specific properties of the lognormal distribution. In this section we give an instructive example of the impact of the lognormal assumption on convexity correction calculations.

Convexity correction arises from second order moments like $\mathbb{E}(y^2)$ or $\mathbb{E}(yR)$. Let us concentrate on second order moments of the form $\mathbb{E}(y^2)$ as the same result can also be derived for expressions of the form $\mathbb{E}(yR)$ but the mathematics are slightly more complicated.

For a lognormal random variable $y(T)$ we have the well known result $\mathbb{E}(y(T)^2) = \bar{y}^2 e^{\sigma^2 T}$, where $\bar{y} = \mathbb{E}(y(T))$ and $\sigma^2 T$ denotes the variance of $\log y(T)$. We want to investigate the contribution of the tail of the integral from U to ∞. In the case of a lognormal random variable we can calculate this value analytically as

$$\int_U^\infty \bar{y}^2 \exp\{-\sigma^2 T + 2\sigma\sqrt{T}\epsilon\} \frac{e^{-\frac{1}{2}\epsilon^2}}{\sqrt{2\pi}} \, d\epsilon = \bar{y}^2 e^{\sigma^2 T} N(-U + 2\sigma\sqrt{T}). \quad (11.38)$$

Often when one thinks about probability distributions, the "relevant" part of the distribution is located within the 95% confidence interval. For the standard normal variable the 95% confidence interval is approximately given by $[-2, 2]$, which implies for the lognormal variable a confidence interval of $[\bar{y}e^{-2\sigma\sqrt{T}}, \bar{y}e^{2\sigma\sqrt{T}}]$. Hence, the contribution to the integral of the values above $U = 2$ is equal to

$$\bar{y}^2 e^{\sigma^2 T} N(-2 + 2\sigma\sqrt{T}). \quad (11.39)$$

For small values of $\sigma\sqrt{T}$ the term $N(-2 + 2\sigma\sqrt{T})$ is indeed small. However, for very long maturities the contribution of the tail can become significant. If we take for example $\sigma = 10\%$ and $T = 50$ then $2\sigma\sqrt{T} = 1.41$ and $N(-2 + 1.41) = 0.28$, which implies that the tail of the distribution outside the 95% confidence interval does contribute more than 25% to the level of the convexity correction. Although this is quite an extreme example, it does show that since convexity corrections are determined by higher order moments, (implicit or explicit) assumptions made about the shape of the tail of the probability distribution can have a significant impact of the magnitude of the convexity correction especially for long maturities. It is also clear that third order (and higher order) moments which are ignored in the derivation of convexity correction formulæ can become more important at very long maturities.

Note also that many derivations of convexity corrections involve approximations which are often justified on the basis of Taylor approximation arguments. For small values of $\sigma\sqrt{T}$ (i.e. short maturities) these approximations are often quite accurate. However, for long maturities the accuracy of the approximations may deteriorate rapidly. For example, when pricing a 30-year CMS swap indexed on the 30Y rate (these products have actually been traded) one should be very careful and consider the assumptions and approximations made by the convexity correction formula in use.

11.6 Appendix: Linear Swap Rate Model

In this appendix we derive the Linear Swap Rate Model (LSM), which is used extensively throughout this chapter to calculate convexity corrections. The LSM is due to Hunt and Kennedy (2000, Chapter 13). Let $y(t)$ be a forward par swap rate with swap start date T_0, $P(t) = \sum_1^N \alpha_{i-1} D_i(t)$ the PVBP of the swap and $\mathbb{Q}^P$ the martingale measure associated with the numeraire P. The swap rate $y(t)$ is defined as $y(t) = (D_0(t) - D_N(t))/P(t)$ and is therefore a martingale under $\mathbb{Q}^P$. Define $\hat{D}_S(t) = D_S(t)/P(t)$, then $\hat{D}_S(t)$ is also a martingale under the measure $\mathbb{Q}^P$.

In the LSM the assumption is made that $\hat{D}_S(T_0) = A + B_S y(T_0)$, where A is a constant and B_S is a deterministic function of S. Hence, it is assumed that the PVBP rebased discount factor $\hat{D}_S$ can be approximated by a linear expression in the swap rate y. Although this is a crude assumption, it can partly be justified by interpreting the LSM as a first order Taylor approximation to any one-factor model driven by the swap rate y. A compelling advantage of the LSM is its analytical tractability.

Let us solve for the parameter A and the function B_S for $S \geq T_0$. We know that $\hat{D}_S$ should be a martingale. To check the martingale property we consider

$$\hat{D}_S(0) = \frac{D_S(0)}{P(0)} = \mathbb{E}^P\left(\hat{D}_S(T_0)\right) = \mathbb{E}^P\left(A + B_S y(T_0)\right) = A + B_S y(0). \quad (11.40)$$

The martingale property of $\hat{D}_S$ is ensured if we set

$$B_S = \frac{D_S(0)/P(0) - A}{y(0)}. \quad (11.41)$$

The parameter A can be determined as follows. Consider the identity $1 \equiv \sum_1^N \alpha_{i-1} \hat{D}_i(t)$. Expanding this expression yields

$$1 \equiv \underbrace{\left(A \sum_{i=1}^N \alpha_{i-1}\right)}_{1} + \underbrace{\left(\sum_{i=1}^N \alpha_{i-1} B_{T_i}\right)}_{0} y(t). \quad (11.42)$$

Hence, we find $A = 1/\sum_1^N \alpha_{i-1}$. It is left as an exercise to the reader to check the condition on the B_{T_i}.

Note that, if y is a forward LIBOR rate L_i, then we have that $P(t) = \alpha_i D_{i+1}(t)$. The LSM model now generates the following expression

$$\hat{D}_i(t) = \frac{D_i(t)}{\alpha_i D_{i+1}(t)} = A + B_{T_i} L_i = \frac{1}{\alpha_i} + L_i(t), \qquad (11.43)$$

which reflects the exact definition of the forward LIBOR rate L_i.

12. Extensions and Further Developments

In the final chapter of this book I would like to discuss extensions of the models presented here. In this chapter I will express my personal opinion and experience on working with interest rate models and how to adapt and extend these models for various purposes. Note that this final chapter is written in the "I" form to emphasise the fact that I express my personal views here. I feel this is necessary, as the practical implementation of pricing models is as much an art as it is pure science.

12.1 General Philosophy

My general philosophy is that the models discussed in this book so far are extrapolation tools that help us to determine the value of illiquid interest rate derivatives on the basis of information we have extracted out of the prices of liquidly traded interest rate derivatives. This is the sole purpose these models are designed for, and this ought to be the only purpose these models are used for. Hence, these models are not necessarily designed to make predictions about future levels of interest rates nor to give a realistic description about the economy. (This does not mean we place no value on economic "realism" as models with realistic properties are probably better extrapolation tools than models with unrealistic properties.)

Since the liquidly traded interest rate derivatives are European-style caps/floors and swaptions, the information we can extract from these prices are the marginal probability distributions of individual LIBOR or swap rates. Most exotic interest rate derivatives have a value that depends on multiple interest rates. Hence, to price these derivatives we need the joint probability distribution of the relevant interest rates. There are many ways to construct a joint probability distribution which is consistent with a given set of marginal probability distributions for the individual interest rates. The essence of what any of the models discussed in this book do, is to choose a particular method of constructing the joint probability distribution. Hence, a model for pricing derivatives "extrapolates" a joint probability distribution on the basis of the marginal probability distributions implied by the prices of the vanilla instruments.

Let me give an example. The Markov-Functional model of Chapter 9 is capable of exactly fitting the marginal probability distributions of a given set of LIBOR or swap rates. The construction of the joint probability distribution of these rate is done implicitly on the basis of the assumption that there exists an underlying Markov process driving the economy and the choice of the law of the underlying Markov process. Note that for any choice of the law of the underlying Markov process we obtain a Markov-Functional model that is consistent with the (marginal) probability distributions of the LIBOR or swap rates but has a different implied joint probability distribution and will therefore very likely calculate a different value for a given exotic option. Because the joint probability distribution lives in a high-dimensional space (one dimension for each interest rate) it is not easy to develop an intuitive feel for the shape of the joint distribution. For this reason we have restricted the choice of possible laws of the underlying Markov process to a set of Gaussian laws parametrised by a single parameter which is the mean-reversion parameter. Thinking about the different underlying processes in terms of mean-reversion makes it easier to develop an intuitive feel for the impact on the prices of exotic options.

The importance of mean-reversion (i.e. the importance of the way the joint distribution is constructed) is illustrated by Table 9.5 of Chapter 9. The table shows that the impact of different levels of mean reversion is far greater than the differences in the marginal probability distributions implied by the different models.

12.2 Multi-Factor Models

An obvious way to extend the models describe in this book is to use multiple sources of uncertainty that drive the economy. This leads to multi-factor interest rate models. From the extrapolation point of view given above, a multi-factor model is just a more elaborate way of constructing joint probability distributions for the relevant interest rates. As a multi-factor model can provide a more realistic description of the economy, the joint distributions implied by multi-factor models may be of better quality but this is not necessarily so. Because the stochastic processes implied by multi-factor models are more complicated, it is even more difficult to fathom the impact of various assumptions on the joint probability distributions implied by the model. As stressed by Rebonato (1998, 1999) a carefully calibrated multi-factor model can still be a very poor pricing tool for some products.

Every model is only a simplification of the real world to help us understand a specific aspect of the real world. If one is working with very complicated models like the interest rate models described in this book, I am very much in favour of using a model that is as simple as possible for the application at hand. I feel much more comfortable using a simple model for which I can understand its limitations and know when the assumptions that the model makes begin to break down. The application of this approach is illustrated in Chapter 11. By using such a product based modelling approach, I can focus my modelling effort on the relevant interest rates and the assumptions and approximations made become much more apparent. Hence for each new product, I force myself to think about the most appropriate model and I force myself to think about the assumptions and approximations needed to derive the pricing model.

This in contrast to the use of a "big" multi-factor model that is used for the valuation of many different types of products. It is very appealing to build such a model as a general pricing framework, which then allows for a quick implementation of new products. Although potentially a very powerful approach, the danger is very real that (because of the ease with which new products can be added) one no longer questions the validity of the assumptions of the model. Also if market conditions start changing, the hidden assumptions made by the model may have unexpected consequences for some products.

A point in case is that several banks lost money during 1999 because the stable relation between caplet and swaptions volatilities broke down. Rumour had it that these losses could be attributed to the fact that the pricing models in use by those houses relied implicitly on a fixed relation between caplet and swaption volatilities.

12.3 Volatility Skews

Another way to extend the market models described in Part II of this book is to relax the assumption that the interest rates have a lognormal distribution. Because the market is so used to using the Black model for pricing European-style options, this is reflected in the market by the fact that a different volatility is used to price options at different strikes and this is called a *volatility skew* by market practitioners. Suppose we denote the implied volatility at different strikes K to price options with a given maturity by the function $\sigma(K)$. Then the price of a European option is calculated using the Black formula as $B(\sigma(K), K)$.

Mathematically speaking, using a volatility skew implies that a different probability distribution than the lognormal distribution is used to calculate the value of an option. Given the prices $B(\sigma(K), K)$ across a range of strikes for caplets or swaptions it is very easy to fit the Markov-Functional model of Chapter 9 to these prices. We refer to Part Two of Rebonato (1999) for a good overview of other methods on how to extract these implied distributions from an observed volatility smile $\sigma(K)$.

However, the same warning about hidden assumptions is appropriate here. By "throwing away" the lognormal assumption, many of the analytical results we take for granted disappear as well. For example, the value of a European digital options is given by the derivative with respect to the strike of a European call option. In case of a volatility smile $\sigma(K)$ we find for the value

$$V^{\text{DIG}}(0) = \frac{\partial B(\sigma(K), K)}{\partial K} = D(0)N(d_2) + \frac{\partial B(\sigma(K), K)}{\partial \sigma} \frac{\partial \sigma(K)}{\partial K}. \quad (12.1)$$

Note that this formula consists of the familiar $N(d_2)$ term from the Black formula (evaluated with volatility $\sigma(K)$) *plus* an additional term proportional to the vega of the call option to correct for the fact that the underlying rate has no longer a lognormal distribution.

Another example are the convexity correction formulæ derived in Chapter 11. The analytical expressions derived have terms of the form $\exp\{\sigma^2 T\}$. These terms are generated by second order moments of the form $\mathbb{E}(y^2)$. Only for the lognormal distribution these moments evaluate to the familiar $\exp\{\sigma^2 T\}$ terms. Other probability distributions have different expressions for the second order moment. Indeed many observed volatility skews reflect the fact that the underlying rates have probability distributions with fat tails, but this can have a significant impact on the magnitude of the convexity correction. See also the discussion in Chapter 11, Section 11.5.

References

Arnold,L. (1992): *Stochastic Differential Equations: Theory and Applications*. Reprint Edition, Krieger Publishing Company, Florida.

Baxter,M. and A. Rennie (1996): *Financial Calculus: An Introduction to Derivative Pricing*. Cambridge Univeristy Press, Cambridge.

Beaglehole,D. and M. Tenney (1991): General Solutions of Some Interest Rate-Contingent Claim Pricing Equations. *Journal of Fixed Income*, Vol. 1, No. 2, pp. 69–83.

Black,F. (1976): The pricing of Commodity Contracts. *Journal of Financial Economics*, Vol. 3, pp. 167–179.

Black,F., E. Derman and W. Toy (1990): A One-Factor Model of Interest Rates and its Applications to Treasury Bond Options. *Financial Analysts Journal*, January-February 1990, pp. 33–39.

Black,F. and P. Karasinski (1991): Bond and Option Pricing when Short Rates are Lognormal. *Financial Analysts Journal*, July–August, pp. 52–59.

Black,F. and M. Scholes (1973): The Pricing of Options and Corporate Liabilities. *Journal of Political Economy*, Vol. 3, pp. 637–654.

Brace,A. (1998): Simulation in the GHJM and LFM Models. FMMA Notes.

Brace,A., D. Gatarek and M. Musiela (1997): The Market Model of Interest Rate Dynamics. *Mathematical Finance*. Vol. 7, pp. 127–154.

Brown,S. and P. Dybvig (1986): The Empirical Implications of the Cox, Ingersoll, Ross Theory of the Term Structure of Interest Rates. *The Journal of Finance*, Vol. 41, pp. 617–630.

Chan,K., A. Karoly, F. Longstaff and A. Sanders (1992): An Empirical Comparison of Alternative Models of the Short-Term Interest Rate. *The Journal of Finance*, Vol. 47, pp. 1209–1227.

Cox,J., J. Ingersoll and S. Ross (1985): A Theory of the Term Structure of Interest Rates. *Econometrica*, Vol. 53, No. 2, pp. 385–407.

Davidson,R. and J. MacKinnon (1993): *Estimation and Inference in Econometrics*. Oxford University Press, Oxford.

De Munnik,J. and P. Schotman (1994): Cross Section versus Time Series Estimation of Term Structure Models. *Journal of Banking and Finance*, Vol. 18, pp. 997–1025.

De Jong,F. (1999): Time-Series and Cross-Section Information in Affine Term Structure Models. *Journal of Economics and Business Statistics*, forthcoming.

De Jong,F., J. Driessen and A. Pelsser (1999): LIBOR and Swap Market Models for the Pricing of Interest Rate Derivatives: An Empirical Analysis. Working paper, University of Amsterdam.

Duffie,D. (1994): *Transform Methods for Solving Partial Differential Equations*, CRC Press, Boca Raton, Florida.

Dupire,B. (1994): Pricing with a Smile. *Risk*. Vol. 9(3), pp. 18–20.

Fisher,G. and M. McAleer (1981): Alternative Procedures and Associated Tests of Significance for Non-Nested Hypotheses. *Journal of Econometrics*, Vol. 16, pp. 103–119.

Garman,M. and S. Kohlhagen (1983): Foreign Currency Option Values. *Journal of International Money and Finance*, Vol. 2, pp. 231–237.

Geman,H., N. El Karoui and J. Rochet (1995): Changes of Numéraire, Changes of Probability Measure and Option Pricing. *Journal of Applied Probability*, Vol. 32, pp. 443–458.

Griffel,D. (1993): *Applied Functional Analysis*, Ellis Horwood Ltd., Chichester.

Harrison,J.M. and D. Kreps (1979): Martingales and Arbitrage in Multiperiod Securities Markets. *Journal of Economic Theory*, Vol. 20, pp. 381–408.

Harrison,J.M. and S. Pliska (1981): Martingales and Stochastic Integrals in the Theory of Continuous Trading. *Stochastic Processes and their Applications*, Vol. 11, pp. 215–260.

Heath,D., R. Jarrow and A. Morton (1992): Bond Pricing and the Term Structure of Interest Rates: A New Methodology for Contingent Claims Valuation. *Econometrica*, Vol. 60, No. 1, pp. 77–105.

Heath,D., R. Jarrow and A. Morton (1996): Bond Pricing and the Term Structure of Interest Rates: A Discrete Time Approximation. *Journal of Financial and Quantitative Analysis*, Vol. 25, pp. 419–440.

Heston, S. (1993): A Closed-Form Solution for Options with Stochastic Volatility with Applications to Bond and Currency Options. *The Review of Financial Studies*, Vol. 6 (2), pp. 327–343.

Ho,T., and S. Lee (1986): Term Structure Movements and Pricing Interest Rate Contingent Claims. *The Journal of Finance*, Vol. 41, pp. 1011–1029.

Hochstadt,H. (1973): *Integral Equations*, John Wiley & Sons, New York.

Hull,J. (2000): *Options, Futures, and Other Derivative Securities* (Fourth Edition): Prentice-Hall, Englewood Cliffs, New Jersey.

Hull,J. and A. White (1990a): Pricing Interest-Rate-Derivative Securities. *The Review of Financial Studies*, Vol. 3 (4), pp. 573–592.

Hull,J. and A. White (1990b): Valuing Derivative Securities Using the Explicit Finite Difference Method. *Journal of Financial and Quantitative Analysis*, Vol. 25 (1), pp. 87–100.

Hull,J. and A. White (1994): Numerical Procedures for Implementing Term Structure Models I: Single-Factor Models. *The Journal of Derivatives*, Fall 1994, pp. 7–16.

Hunt,P. and J. Kennedy (2000): *Financial Derivatives in Theory and Practice*. John Wiley & Sons, Chichester.

Hunt, P., J. Kennedy and A. Pelsser (2000): Markov-Functional Interest Rate Models. *Finance and Stochastics*, forthcoming.

Hunt, P. and A. Pelsser (1998): Arbitrage-Free Pricing of Quanto-Swaptions. *Journal of Financial Engineering*, Vol. 7(1), pp. 25–33.

Jamshidian,F. (1989): An Exact Bond Option Formula. *Journal of Finance*, Vol. 44, pp. 205–209.

Jamshidian,F. (1991): Bond and Option Evaluation in the Gaussian Interest Rate Model. *Research in Finance*, Vol. 9, 131–170.

Jamshidian,F. (1995): A Simple Class of Square-Root Interest Rate Models. *Applied Mathematical Finance*. Vol. 2, pp. 61–72.

Jamshidian,F. (1996): Bond, Futures and Option Evaluation in the Quadratic Interest Rate Model. *Applied Mathematical Finance*, Vol. 3, pp. 93–115.

Jamshidian,F. (1998): LIBOR and Swap Market Models and Measures. *Finance and Stochastics*. Vol. 1(4), pp. 293–330.

Judge,G., C. Hill, W. Griffiths, H. Lütkepohl, T.-C. Lee (1982): *Introduction to the Theory and Practice of Econometrics.* John Wiley & Sons, New York.

Karatzas,I. and S. Shreve (1991): *Brownian Motion and Stochastic Calculus* (Second Edition). Springer Verlag, Berlin.

Karatzas,I. and S. Shreve (1998): *Methods of Mathematical Finance.* Springer Verlag, Berlin.

Longstaff,F. and E. Schwartz (1992): Interest Rate Volatility and the Term Structure: A Two-Factor General Equilibrium Model. *The Journal of Finance,* Vol. 47, pp. 1259–1282.

Longstaff,F., P. Santa-Clara and E. Schwartz (1999): Throwing Away a Billion Dollars: The Cost of Suboptimal Exercise Strategies in the Swaptions Market. Working paper UCLA.

Lukacs,E. (1970): *Characteristic Functions* (2nd Edition). Charles Griffin & Co., London.

Margrabe,W. (1978): The Value of an Option to Exchange One Asset for Another. *Journal of Finance,* Vol. 33, pp. 177–186.

Miltersen,K., K. Sandmann and D. Sondermann (1997): Closed Form Solutions for Term Structure Derivatives with Lognormal Interest Rates. *Journal of Finance.* Vol. 52, pp. 409–430.

Mizon,G. and J.-F. Richard (1986): The Encompassing Principle and its Application to Testing Non-Nested Hypotheses. *Econometrica,* Vol. 54, pp. 657–678.

Moraleda,J. and A. Pelsser (2000): Forward versus Spot Interest-Rate Models of the Term Structure: An Empirical Comparison. *The Journal of Derivatives,* Vol. 7 (3), pp. 9–21.

Musiela,M. and M. Rutkowski (1997): *Martingale Methods in Financial Modelling.* Springer Verlag, Berlin.

Øksendal,B. (1998): *Stochastic Differential Equations* (Fifth Edition). Springer Verlag, Berlin.

Pedersen,M. (1999): Bermudan Swaptions in the LIBOR Market Model. Working paper Simcorp A/S.

Pedersen,M. and J. Sidenius (1997): Valuation of Flexible Caps. Working paper Skandinavia Enskilda Banken, Copenhagen.

Pelsser,A. (1997). A Tractable Interest Rate Model that Guarantees Positive Interest Rates. *Review of Derivatives Research,* Vol. 1, pp. 269–284.

Pelsser,A. (2000). Convexity Correction. Working paper, Erasmus University Rotterdam.

Press, W., S. Teukolsky, W. Vetterling and B. Flannery (1992): *Numerical Recipes in C: The Art of Scientific Computing* (Second Edition). Cambridge University Press, Cambridge.

Rebonato,R. (1998): *Interest Rate Option Models* (Second Edition). John Wiley & Sons, Chichester.

Rebonato,R. (1999): *Volatility and Correlation in the Pricing of Equity, FX and Interest-Rate Options.* John Wiley & Sons, Chichester.

Reiner,E. (1992): Quanto Mechanics. *Risk,* March, pp. 59–63.

Rogers,L. (1995): Which Model for Term-Structure of Interest Rates Should One Use? In: *IMA Vol. 65: Mathematical Finance,* eds. M. Davis et al., Springer Verlag, Berlin.

Rogers,L. and D. Williams (1994): *Diffusions, Markov Processes and Martingales* (Second Edition), John Wiley & Sons, Chichester.

Smith,G. (1985): *Numerical Solution of Partial Differential Equations: Finite Difference Methods,* (3rd Edition), Oxford University Press, Oxford.

Vasicek,O. (1977): An Equilibrium Characterisation of the Term Structure. *Journal of Financial Economics,* Vol. 5, pp. 177–188.

Wilmott,P. (1998): *Derivatives*. John Wiley & Sons, Chichester.

Wilmott,P., J. Dewynne and S. Howison (1993): *Option Pricing: Mathematical Models and Computation*, Oxford Financial Press, Oxford.

Williams,W. (1980): *Partial Differential Equations*, Clarendon Press, Oxford.

Index

σ-algebra 6

accrual factor 88, 93
accrual period 147
American-style option 16, 51, 57, 64,
  66, 105
  see also early exercise
arbitrage-free 13, 14, 29, 35, 89, 93,
  141, 142
arbitrage-free economy 140
arbitrage opportunity 5–7, 9, 14, 24,
  28, 64
arbitrage pricing 8
Arnold 40, 48
Arrow-Debreu security 35
artificial compound model 76, 82
artificial nesting 75
asset 8, 110, 121
asymptotic 75, 76, 80, 82
attainable 7, 9
auto-cap/floor 109, 125–127
auto-correlation 109, 120, 121, 136

backward induction 18, 55, 66, 113,
  121, 122, 127
barrier (option) 100–101
– cap 109, 121
– digital cap 103
– outside barrier swaption 153
basispoint 20, 57, 66, 74, 101, 123
  see also present value of a basispoint
Baxter 6
Beaglehole 59
Bermudan-style 105, 125, 127
– swaption 105, 109, 127, 128
Black 12, 16, 40, 41, 73, 87, 90, 94, 111,
  128
Black formula 87, 112–114, 120, 140,
  146, 149, 161, 162
Black-Scholes
– economy 12, 13, 16, 29

– formula 13, 18
bond 46, 71, 74, 103
  see also counpon bearing bond
  see also discount bond
boundary condition 17, 18, 25, 34, 36,
  38, 39, 41, 48, 49, 55, 61, 63, 68, 106
Brace 87, 132
Brown 74
Brownian motion 6, 11, 19, 38, 89, 91,
  93, 97, 105, 106, 110, 121
  see also geometric Brownian motion

call 50, 51, 103, 119, 145–147, 162
– option 13, 50, 63, 64
– option on discount bond 63
cap 45, 51, 57, 59, 71, 73–75, 77, 82, 84,
  87, 100, 101, 159
caplet 57, 58, 89, 90, 98, 100, 111, 119,
  122, 125, 126, 131, 136, 161
central limit theorem 20
Chan 74, 84, 133
change of measure 11, 29, 90, 95, 143,
  144
change of numeraire 11, 32, 33, 38, 90,
  140–145
characteristic function 36, 37, 39, 50,
  63
chi-square distribution 60, 72, 82
Cholesky decomposition 107
chooser-cap 109, 125–127
– digital 125
complete economy 7, 8, 13, 141
conditional expectation 35, 110–111
Constant Maturity Swap 149
constant volatility
– LIBOR market model 133, 134, 136
– swap market model 135
continuous trading 77
continuous trading economy 6–9, 16
continuously compounded 57, 77

convexity correction   140, 142–144,
    147–150, 152, 155, 156, 162
correlation   33, 96, 106, 120, 121, 124,
    128, 134, 139, 143–145, 150, 151, 154
– matrix   107
    see also auto-correlation
    see also cross-correlation
coupon bearing bond   51, 59
coupon payment   103
Cox   72, 74
cross currency   144, 145
cross-correlation   120, 136
cross-section   133, 136
cumulative normal distribution function
    13, 50, 59, 60, 64, 72, 112

Datastream   77
date-roll convention   89
Davidson   71, 76
daycount convention   57, 88
daycount fraction   57, 88, 89, 101, 103,
    126, 147, 149
De Jong   71, 131, 133, 136
De Munnik   74, 133
delta-function   34–36
derivative security   7, 8
Derman   73
Dewynne   16, 18
diff option   15
diffed CMS   150
diffed LIBOR   150
diffed LIBOR in Arrears   150
diff(erential) swap   150
digital cap   102
digital floor   102
discontinuous sample path   7
discount bond   25–29, 31–33, 35–37,
    39, 41, 42, 45, 48, 50–52, 55, 57–61,
    63–65, 69, 72, 73, 77, 88, 89, 91, 94, 98,
    101, 103, 107, 109–111, 113, 115, 117,
    126, 127, 142
discount factor   89, 94, 98, 99, 104, 106,
    147, 149, 151, 156
discounted payoff   18
discounted value   143
discrete barrier cap/floor   100–102,
    121
– digital   102
– options   124
discrete dividend payment   7
discrete economy   35
domestic asset   141
double barrier   121

doubling strategy   9
Driessen   131, 136
Duffie   36
Dupire   112
Dybvig   74

early exercise   56, 67, 126–128
early notification   127, 128
equivalent martingale measure
    see martingale measure
equivalent probability measure
    see probability measure
Eurodollar   132
European   13, 17, 50, 51, 57, 63, 66
– call option   162
– digital options   162
– option   87, 90, 104, 119, 127–128,
    159, 161
– swaption   128
exchange rate   14, 15, 75, 141, 142, 150
exercise   125–128
exercise decision   128
exercise price   119
exercise strategy   125, 127
exercise time   13
exotic   64, 88, 109, 120, 121, 131, 153,
    159
exotic European   139, 140
exotic option   46, 60, 100, 129, 160
expiry date   93
explicit finite difference   18, 46, 52, 66

factor   134, 136, 139
    see also one-factor
    see also two-factor
    see also multi-factor
Feynman-Kac formula   17, 25–27, 30,
    48, 61
filtration   6
finite difference   18, 46, 51, 53–55, 57,
    66
    see also trinomial
Fisher   75
fit model to
– prices   74, 75, 80, 81, 84, 87, 109, 111,
    114, 118, 119, 131, 133, 134
– probability distribution   109, 139
– term-structure   25–27, 30, 45, 46, 52,
    57, 59, 60, 64, 65, 72–74
fixed
– amount   102
– interest rate   92, 93, 107
– – payment   92, 114, 136, 149

– leg   51, 92, 151
fixing in arrears   147
flex(ible) cap   125
floating accrual   153
floating
– interest rate   51, 150
– – payment   92, 136
– leg   51, 92, 150
floor   45, 51, 57–59, 71, 73–75, 77, 82, 84, 87, 100, 101, 159
floorlet   90, 127
foreign asset   141
foreign-exchange   *see* exchange rate
forward
– exchange rate   144, 145, 150
– interest rate   140, 142
– LIBOR payment   136
– LIBOR process   90
– LIBOR rate   88, 90, 91, 97, 99, 100, 124, 132, 134, 136, 142, 147, 150, 157
– measure   142
– par swap rate   93, 136, 142, 156
– position   127
– price   128
– rate   27–30, 33, 52, 65, 70
– swap rate   91, 95, 104–106, 128, 135, 149
– swap rate process   145
forward rate model   23, 30
forward value   128
Fourier transform   31, 36, 41, 48–50, 61
Fubini's Theorem   26
   see also interchanging order of integration
fundamental solution   31–34, 36, 37, 39–41, 43, 45, 46, 48, 59, 61
futures contract   132
futures/forward correction   132

Garman   14, 15
Gatarek   87
Gaussian   116–118, 121, 160
   see also normal distribution
Geman   11, 31
generalised method of moments   74
geometric Brownian motion   12, 14, 87
Girsanov's Theorem   11, 12, 14, 15, 29, 38, 90, 91, 95
goodness-of-fit criteria   75, 80
Griffel   34, 68

Harrison   8–10
Heath   23, 26, 27, 139

hedge   24
– parameter   115
– ratio   75
Heston   36
heteroscedasticity   74
higher order moment   155
Ho   26
Ho-Lee model   27, 30, 37, 38, 40, 45, 46, 48, 50
Hochstadt   68
Howison   16, 18
Hull   15, 37, 40, 45, 46, 48, 51–53, 55, 57, 72–74, 140, 148–150
Hull-White model   41, 45, 46, 49, 50, 52, 55, 57, 59, 66, 71–73, 111, 120, 121, 128
Hunt   6, 30, 109, 115–117, 128, 151, 156

implied volatility   112, 128, 132–135, 145, 161
– caplet volatility   90
– swaption volatility   94
implied distribution   161
infinitesimal time-period   23
Ingersoll   72, 74
instantaneous correlation   105, 106, 152
instantaneous interest rate   14, 23, 27, 33, 64
instantaneous volatility   143
integral equation   65–68, 111
Intercapital Brokers   71, 77
interchanging order of integration   26, 28, 67
interest rate swap   92
inverse cumulative normal distribution function   113
Itô
– integral   7, 8
– process   6–7
– lemma   12–16, 24, 29, 90, 96, 104, 105

J-test   76
Jamshidian   31, 51, 59, 60, 65, 69, 72, 87
Jarrow   23, 26, 27, 139
joint
– distribution   109, 129, 139, 159–160
   see also marginal distribution
– law   143
Judge   75

Karasinski   40, 41, 73, 128
Karatzas   6, 7, 11
Karoly   74, 84
Kennedy   6, 30, 109, 115–117, 128, 156
knock-out   *see* barrier (option)
– (digital)cap/floor   124
Kohlhagen   14, 15
Kreps   8, 10

least squares   71, 75, 76
Lee   26
LIBOR   57, 84, 87–90
– in Arrears   147
– – caplet   148
LIBOR market model   87, 89, 90, 94,
    95, 97, 99–101, 103, 111, 121, 123, 131,
    132, 154
limit cap/floor   125
linear combination   24, 34
Linear Swap Rate Model   140,
    143–146, 148, 150, 156
local martingale   8, 9
locally riskless   24
lognormal   90, 94, 112, 143, 146, 147,
    155, 161
– distribution   73, 82, 84, 89, 96, 146,
    152, 155, 161, 162
lognormal martingale   93, 112, 144,
    146
lognormal model   71, 73, 84
lognormal process   145
Longstaff   74, 84, 133, 134
Lukacs   36, 63

MacKinnon   71, 76
marginal distribution   127, 129, 139,
    159, 160
    see also joint distribution
Margrabe   146, 153, 154
market model   87, 105, 109, 110, 131,
    133, 142, 154, 161
market price of risk   25, 26, 30, 40, 47,
    61
marketed asset   6–9, 12, 15, 16, 27–29,
    88, 89
Markov   109–112, 115, 116, 118, 120,
    121, 160
Markov-Functional model   20, 109,
    123, 131, 160, 161
martingale   6, 8–10, 13–15, 29, 32, 33,
    89–93, 95, 97, 104, 105, 141–145, 147,
    149, 151, 156

    see also local martingale
    see also lognormal martingale
martingale representation theorem   9
martingale measure   8–16, 18, 23,
    27–32, 37, 47, 61, 89, 109, 110, 118,
    121, 140–144, 147, 156
    see also change of measure
    see also probability measure
    see also T-forward-risk-adjusted
    measure
    see also terminal measure
maturity   25, 27–29, 31, 32, 35, 37, 48,
    50, 55, 61, 63, 69, 74, 88, 107, 117, 119,
    127, 133, 142, 145, 149, 151, 161
McAleer   75
mean-reversion   45, 48, 80, 81, 109,
    120, 121, 123, 124, 128, 129, 131,
    133–136, 154, 160
mean-reversion LMM   133, 134
mean-reversion SMM   135
Miltersen   87
Mizon   75
moment   37, 155
    see also higher order moment
    see also second order moment
money-market account   10, 12–15, 27,
    28, 32, 37, 88
Monte Carlo   16, 20, 87, 91, 97, 100,
    101, 103–105, 121, 124, 135, 154
    see also variance reduction technique
Moraleda   71
Morton   23, 26, 27, 139
multi-factor   7, 23, 60, 72, 96, 105, 106,
    136, 139, 140, 160, 161
Musiela   6, 87, 142

no-arbitrage   8, 16
non-parametric   117
normal distribution   13, 19, 26, 31,
    39–73, 75, 76, 82, 97
normal model   31, 40, 41, 45, 46, 60, 71,
    73
numeraire   8–13, 15, 18, 29, 31, 32, 37,
    89, 91, 93, 103, 107, 110, 111, 113–115,
    117–119, 127, 140–144, 147, 156
– rebased   98–102, 104–106
    see also relative price
numerical integration   16, 20, 65, 115,
    117, 121, 152
    see also Simpson's rule
    see also trapezoid rule

Øksendal   11, 17

one-factor   23, 25, 27, 37, 40, 59, 60, 71, 72, 84, 96, 100, 120, 121, 133, 139, 154, 156
option
– on coupon bearing bond   45, 51
– on a discount bond   45, 50–51, 55, 59–60, 64–65, 72–73
option pricing   5, 6, 12, 16
ordinary differential equation   28, 39, 49, 60, 62, 68
Ornstein-Uhlenbeck process   48, 73

P-test   71, 76, 77, 82, 84
par swap rate   92–96, 114, 132
parsimonious   75
partial differential equation   16–18, 23, 25, 26, 31–34, 36–43, 46, 47, 53, 54, 60, 61, 66
payer swap   92, 93
payer swaption   93, 94
payment stream   103
Pedersen   105, 125
Pelsser   59, 71, 109, 128, 131, 136, 139, 151
Pliska   8, 9
present value of a basispoint   93
Press   20, 107, 115, 116
principal component   139
probability measure   6, 8, 9, 29, 32, 33, 38, 90, 94, 140–142
put   51, 64, 103, 145
– option   51, 57, 58, 66

quanto
– option   15
– swaptions   151
– CMS   150
– LIBOR   150

Radon-Nikodym derivative   11, 33, 37, 38, 90, 141, 143–146
    see also change of measure
    see also change of numeraire
ratchet option   103
Rebonato   20, 134, 139, 160, 161
receiver swap   92, 93
receiver swaption   93, 94
Reiner   15
relative price   8, 9, 12–15, 29, 127
    see also numeraire rebased
– discount bond price   119
Rennie   6
replicating portfolio   8, 16

replicating trading strategy   7–10
Richard   75
Richardson   71
risk-neutral economy   61
Rogers   17, 72, 74, 84
Ross   72, 74
Rutkowski   6, 142

Sanders   74, 84
Sandmann   87
Santa-Clara   133, 134
Scholes   12, 16
Schotman   74, 133
Schwartz   74, 133, 134
second order moment   155, 162
self-financing trading strategy   7, 10, 16
semi-annual compounding   149
semi-parametric   118, 119
Shreve   6, 7, 11
Sidenius   125
Simpson's rule   115
skewed distribution   82, 84
– skewed to the right   71, 82, 84
Smith   53
Sondermann   87
source of uncertainty   6, 7, 23, 160
spot rate
– exchange rate   150
– interest rate   23, 24, 27, 28, 30, 31, 33, 37–40, 45–47, 49, 52, 57, 59–62, 64, 66, 72, 73, 87, 110, 120
– LIBOR rate   88, 97, 100, 132, 150, 153
spot rate model   18, 23, 30, 71, 84, 87, 109, 110, 131
spread
– cap/floor   154
– option   106, 107, 151, 153, 154
– rate based   106, 107, 151
– value based   106, 107
square root model   72
squared Gaussian model   37, 40, 59, 60, 62, 64, 66, 71–73
standard error   20, 75, 77, 80, 82, 84, 101, 123
sticky cap   103
stochastic calculus   11, 12
stochastic differential equation   6, 12, 13, 16, 18, 19, 23, 26, 28, 38, 40, 72, 89, 91, 98
stochastic integral   8, 9, 26
stochastic integral equation   18

stochastic process   9, 11, 12, 24, 28, 30,
   31, 48, 74, 84, 139, 160
stochastic variables   31
stochastic volatility   36
   see also volatility skew/smile
strike   13, 51, 57, 58, 74, 77, 100–103,
   111, 117, 118, 122, 151, 154, 161, 162
swap   51, 71, 77, 92, 93
– swap rate   84, 87, 94
– swap tenor   92
– swaption   45, 51, 87, 93, 111, 159
swap market model   87, 88, 91, 93–97,
   104, 106, 107, 131

T-forward-risk-adjusted measure   31,
   33, 35, 37, 39, 46, 48, 50, 52, 61, 63–65,
   69
T_1-forward measure   95, 96, 105, 106
tail   80, 155, 162
Taylor approximation   76, 155, 156
Tenney   59
tenor   88, 93, 110, 111, 132
term-structure of interest rates   23, 25,
   30, 52, 53, 57, 64, 66, 71, 74, 119, 127,
   132, 149
– LIBOR rates   99
– swap rates   104
term-structure of volatilities   71, 99,
   104
– caplet volatilities   126
terminal measure   91, 94, 95, 97, 100,
   104, 105, 113, 114
total maturity   132, 133, 135
Toy   73
traded asset   141, 143
trading strategy   6–8
transaction costs   5, 7

transformation of variables   47, 48, 53,
   60, 73
trapezoid rule   115
trigger   153
– caplet   153
– swaption   153
trinomial
– algorithm   66
– probability   54
– tree   55, 59, 60, 64, 66, 73, 80
   see also finite difference
two-factor   106, 136, 154

underlying Markov process   160
underlying process   31, 32, 40, 45, 46,
   49, 59, 60, 62, 66, 73, 160

variance reduction technique   20, 101
Vasicek   23, 40
volatility   12, 48, 73, 112, 114, 120, 124,
   128, 131–136, 144–146, 148–150, 152,
   154, 161
   see also term-structure of volatility
   see also stochastic volatility
– skew/smile   161, 162
Volterra integral equation   67, 68

White   37, 40, 45, 46, 48, 51–53, 55, 57,
   72–74
Williams   17, 34
Wilmott   16, 18

yield-curve model   25, 40, 64, 73–75
   see also spot rate model

zero-curve   57, 66, 77, 101, 126
   see also term-structure of interest
   rates